Advanced Fossil Energy Utilization 2014

Topical Conference at the 2014 AIChE Annual Meeting

Atlanta, Georgia, USA
16-21 November 2014

ISBN: 978-1-5108-1250-5

Printed from e-media with permission by:

Curran Associates, Inc.
57 Morehouse Lane
Red Hook, NY 12571

Some format issues inherent in the e-media version may also appear in this print version.

Printed by Curran Associates, Inc. (2015)

AIChE
120 Wall Street, FL 23
New York, NY 10005-4020

Phone: (800) 242-4363
Fax: (203) 775-5177

www.aiche.org

Additional copies of this publication are available from:

Curran Associates, Inc.
57 Morehouse Lane
Red Hook, NY 12571 USA
Phone: 845-758-0400
Fax: 845-758-2634
Email: curran@proceedings.com
Web: www.proceedings.com

TABLE OF CONTENTS

The SECARB Integrated CO2 Capture, Transportation and Storage Project

Gerald Hill, Southern States Energy Board; Richard Esposito, Southern Company; Richard Rhudy and Robert Trautz, Electric Power Research Institute, and George Koperna, Advanced Resources International.

The United States Department of Energy (DOE) established seven Regional Carbon Sequestration Partnerships (RCSPs). The Southeast Regional Carbon Sequestration Partnership (SECARB), whose lead organization is the Southern States Energy Board (SSEB), represents 13 States within the south eastern United States.

The SECARB Anthropogenic Test R&D project is a pilot scale deployment of CO_2 capture, transportation, geologic storage and monitoring technology. The capture component of the test takes place at the James M. Barry Electric Generating Plant (Plant Barry) in Bucks, Alabama. The capture facility, equivalent to 25 megawatts (MW), utilizes post-combustion amine capture technology licensed by Mitsubishi Heavy Industries America (MHIA). CO_2 captured at the plant is transported by pipeline for underground storage in a deep, saline geologic formation within the Citronelle Dome located in Citronelle, Alabama. The injection wells and surface facilities are located on a property which is owned fee simple by the field's operator, Denbury Onshore, LLC.

Since August 2012, up to 550 metric tonnes of CO_2 per day, the equivalent emissions from 25 MW of the plant's capacity have been captured for geologic storage. Transportation and injection operations will continue until the fall of 2014. This project is one of the first and the largest fully-integrated commercial prototype coal-fired carbon capture and storage projects in the USA.

Capture Technology. The technology deployed for capturing CO_2 from the power plant is the MHIA KM-CDR process, which utilizes their proprietary KS-1 solvent to achieve high levels of CO_2 retention with significant reductions in energy penalty from current technologies. The CO_2 capture and compression island utilize representative equipment and demonstrate MHIA's approach for process scale-up.

Pipeline Transportation. A 4-inch diameter pipeline was constructed that stretches approximately 19 kilometers (12 miles) from the outlet of the CO_2 capture facility to the point of injection at the Citronelle Field. The fit-for-purpose pipeline was constructed of standard API 5L X-65 grade pipe with wall thickness between 0.48 and 0.56 centimeters. The pipeline was completed and pressure tested (hydrotested) in November 2011.

Injection and Monitoring. The injection target is the lower Cretaceous Paluxy formation which occurs at a depth of 2,865 meters (9,400 feet). Surface and subsurface monitoring was initiated prior to start of injection to establish baseline conditions. Post injection monitoring will conducted from the fall of 2014 through 2017 to track plume movement and monitor for leakage.

387645 Cost Analysis for Retrofit of Coal-Fired Power Plants with Post-Combustion CO2 Capture Technology

Abstract:

The escalating level of atmospheric carbon dioxide is one of the biggest environmental challenges of our age. CO_2 capture and storage from large point sources such as coal-fired power plants is considered a viable path towards CO_2 mitigation. Among the CO_2 capture processes under investigation, post-combustion technology has the advantage that the equipment can be retrofitted to the existing infrastructures. In addition, post combustion capture may be tuned for specific CO_2 capture targets which only require partial capture such as the NSPS proposed limit of 1,100 lbs of CO_2/MWhr for coal fired power plants.

This presentation will discuss the cost analysis for post combustion based capture processes based on the methodology of Case 11, 550 MWe supercritical coal-fired units as discussed in the report by the Department of Energy.[1] The presentation will assess the sensitivity of costs associated with capital and operational expenditures as a result of solvent properties such as energy penalty of the plant equipped with a post-combustion capture technology. Significant current research focuses on reduction of the energy penalty associated with post combustion capture; however, the costs of CO_2 capture may not be dominated by the energy penalty of the process. This presentation will also assess the viability of $40/tonne of CO_2capture targets for post combustion systems and predicted capital cost and energy penalty requirements to achieve DOE targets for 2025.

[1] DOE/NETL 2007, *Cost and Performance Baseline for Fossil Energy Plants, Volume 1: Bituminous Coal and Natural Gas to Electricity,* Rev. 2, DOE/NETL-2010/1397, November 2010.

390890 Performance of Sorption-Enhanced Water-Gas Shift in Integrated Gasification Combined Cycle Using MgO-Based Sorbent

Abstract:

Global warming is a clear and present danger to world's public health, economy, and environment. Coal is the most polluting of all fossil fuels and the largest single source of global warming pollution in the world. Advanced coal-based power generation technologies, such as Integrated Gasification Combined Cycle (IGCC) have been developed to improve the efficiency of power generation and reduce the CO2 emission per until of energy produced. It has been shown that the CO2

temperatures around 350°C, and the CO2 removal can shift WGS reaction to enhance hydrogen production. Therefore, Sorbent Enhanced Water-Gas-Shift (SEWGS) can result in much higher hydrogen production without lowering the temperature, leading to higher overall process efficiency.

This paper presents the experimental results obtained with a highly reactive and attrition resistant regenerable MgO-based sorbent which can simultaneously remove carbon dioxide and enhance hydrogen production in a single packed bed reactor. It was found that the rate of the WGS reaction is dependent on the MgO concentration, which has a significant catalytic affect on promoting the WGS reaction in the temperature range of 300-430°C. A kinetic model was developed to fit the WGS reactivity curves obtained from the packed bed tests at different operating conditions.

in the coal gas can be removed by regenerable MgO-based sorbents at

Analysis of membrane-based processes for CO_2 capture from natural gas combined cycle power plants

Joseph A. Swisher, Abhoyjit S. Bhown
Electric Power Research Institute
Palo Alto, CA 94304

A variety of separation processes have been advanced as candidates for post-combustion removal of CO_2 from fossil-fired electricity generating units. Carbon capture and storage (CCS) processes based on selective gas separation membranes provide several practical advantages, including avoiding steam extraction from the electricity generating cycle. Several groups have proposed membrane-based flowsheets to balance trade-offs between permeability and selectivity.[1-4] In addition, combustion of natural gas in natural gas combined cycles (NGCC) plants have contributed a increasing portion of the electricity generated in the US as a result of the recent low price of natural gas and the relatively low cost to build NGCC systems. As more electricity is generated by burning natural gas, reducing CO_2 emissions may require the use of CCS technology on natural-gas derived flue gas streams. We have used process modeling to quantify the impacts that membrane-based CCS processes could have on NGCC plant output. These process models were used to compare different process flowsheets and to help determine the trade-offs between membrane properties (permeance and selectivity) and process variables such as membrane stage sizes and stream pressures. Numerical optimization techniques were used to minimize either the power used by the capture system or the projected cost of CO_2 capture under different process constraints.

References

1. Toy, L.; Kataria, A.; Gupta, R. P. *CO2 Capture Membrane Process for Power Plant Flue Gas*; Final Technical Report DE-NT0005313; RTI International: Research Triangle Park, NC, 2012; p. 76.
2. Bocciardo, D.; Ferrari, M.-C.; Brandani, S. *Energy Procedia* **2013**, *37*, 932.
3. Merkel, T. C.; Wei, X.; He, Z.; White, L. S.; Wijmans, J. G.; Baker, R. W. *Ind. Eng. Chem. Res.* **2013**, *52*, 1150.
4. Ramasubramanian, K.; Verweij, H.; Winston Ho, W. S. *Journal of Membrane Science* **2012**, *421–422*, 299.

Low-Energy Solvents for CO_2 Capture Enabled by a Combination of Enzymes and Vacuum Regeneration

Alan House

AIChE 2014 meeting
Atlanta, GA
November 17, 2014

Notices

- **ACKNOWLEDGEMENT OF GOVERNMENT SUPPORT.** This material is based upon work supported by the Department of Energy under Award Number DE-FE0007741.
- **DISCLAIMER.** This report was prepared as an account of work sponsored by an agency of the United States Government. Neither the United States Government nor any agency thereof, nor any of their employees, makes any warranty, express or implied, or assumes any legal liability or responsibility for the accuracy, completeness, or usefulness of any information, apparatus, product, or process disclosed, or represents that its use would not infringe privately owned rights. Reference herein to any specific commercial product, process, or service by trade name, trademark, manufacturer, or otherwise does not necessarily constitute or imply its endorsement, recommendation, or favoring by the United States Government or any agency thereof. The views and opinions of authors expressed herein do not necessarily state or reflect those of the United States Government or any agency thereof.
-

Agenda

novozymes
Rethink Tomorrow

- Project Overview
  - Carbonic anhydrase for CO_2 capture
  - Partners, budget & objective
- Technology Background
  - Process concept
  - Fundamental mechanism
- Progress and Status
  - Bench-scale system description
  - Parametric test results
  - 500-h run initial test results
- Conclusions & Next Steps

Carbonic anhydrase for CO_2 capture: a select survey of other public projects

novozymes
Rethink Tomorrow

- Codexis
  - DE-AR0000071
  - PE dissolved CA for 4.2 M MDEA and >85°C
  - Stable performance, 60 h test at NCCC

About the NCCC:
- The National Carbon Capture Center provides a slipstream from a 880 MW coal unit for testing post-combustion CO_2 capture technologies.

- Akermin
  - DE-FE0004228
  - Immobilize CA in absorber
  - Stable performance, 116 days at NCCC
- Illinois State Geological Survey, Prairie Research Institute, Yongqi Lu Laboratory
  - DE-FC26-08NT0005498 and others
  - Evaluate process concepts, immobilization options, temp. stability, kinetics, etc.

5

Project Objective

novozymes Rethink Tomorrow

Complete a *bench-scale study* and corresponding full technology assessment to validate the potential in meeting the DOE Program Objectives of a *solvent-based post-combustion carbon dioxide capture* system that integrates

$$CO_2 + H_2O + K_2CO_3 \leftrightarrow 2KHCO_3$$

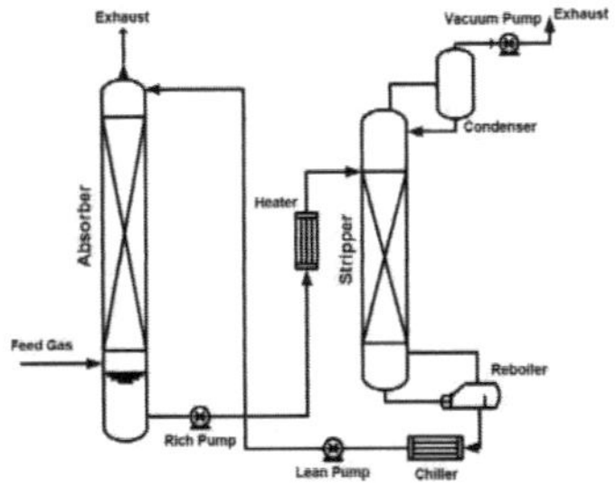

- a **low-enthalpy**, aqueous potassium carbonate-based solvent
- with an **absorption**-enhancing (*dissolved*) carbonic anhydrase enzyme catalyst
- and a low temperature vacuum **regenerator**
- in a **re-circulating** absorption-desorption process configuration

6

Novozymes in Brief – World Leader in Bioinnovation Producing large volume enzymes for industrial applications

novozymes Rethink Tomorrow

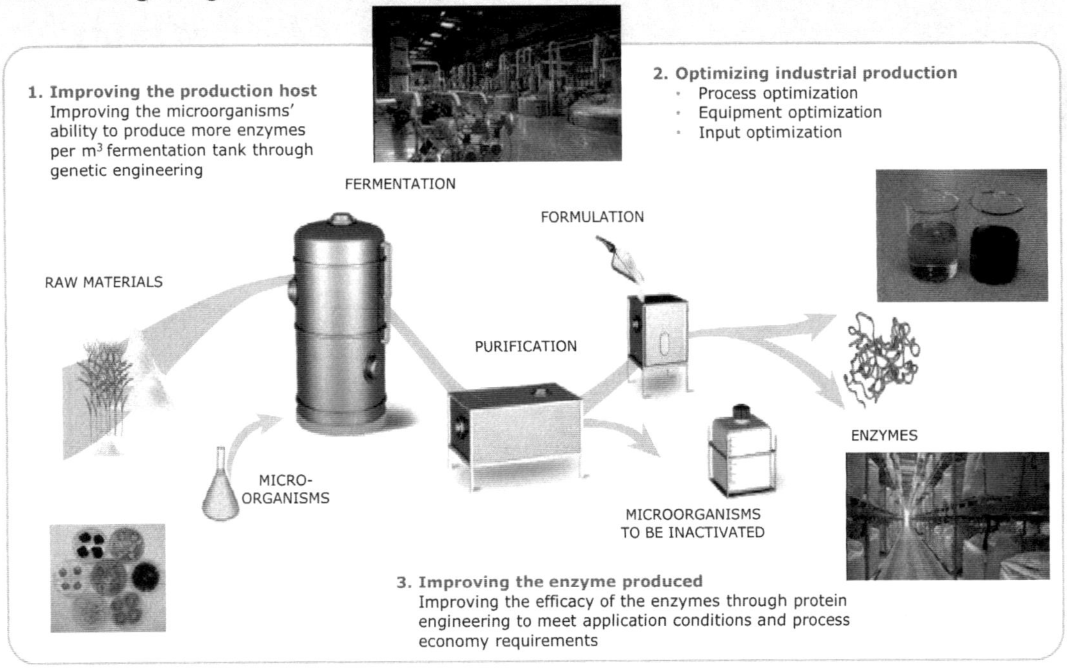

www.novozymes.com

7

Project Overview

novozymes Rethink Tomorrow

- Project Participants

Pacific Northwest National Laboratory	Doosan	novozymes	UK Center for Applied Energy Research
Ultrasonics & Aspen®	Full Process Analysis	Enzymes & Solvents	Kinetics & Bench-scale Tests

- DOE Project Manager: Andrew Jones
- Project Number: DE-FE0007741
- Total Project Budget: $2,088,644
  - DOE: $1,658,620
  - Cost Share: $430,024
- Project Duration: Oct. 1, 2011 – March 31, 2015

DOE Program Objectives
Develop solvent-based, post-combustion technology that
- Can achieve ≥ 90% CO_2 removal from coal-fired power plants
- Demonstrates progress toward the DOE target of <35% increase in LCOE.

8

Process Concept, Advantages & Challenges

novozymes Rethink Tomorrow

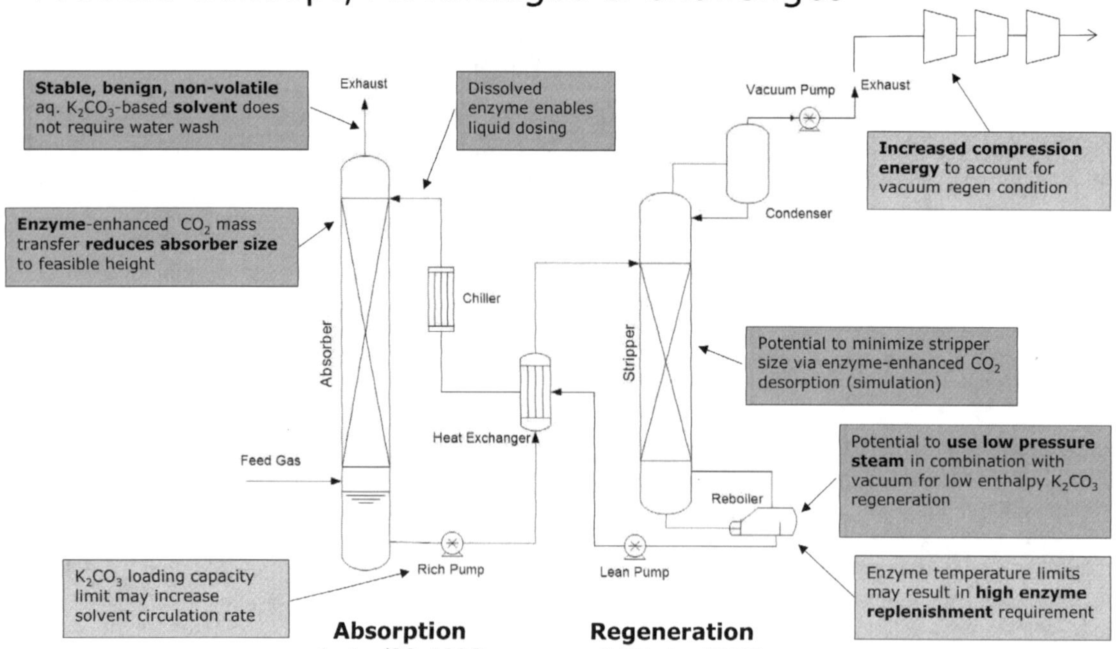

Absorption
1 atm/30-40°C

Regeneration
~0.35 atm/76°C

Generating Bench-scale Test Data

novozymes Rethink Tomorrow

Enzyme Enhanced CO_2 Absorption Mechanism

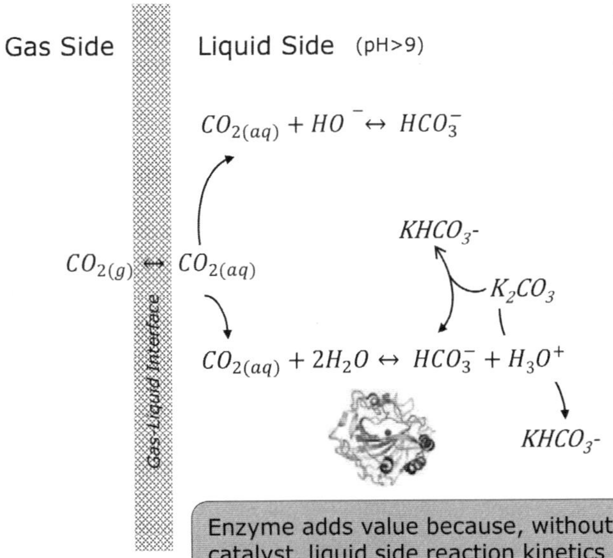

Enzyme adds value because, without catalyst, liquid side reaction kinetics are overall mass transfer rate limiting

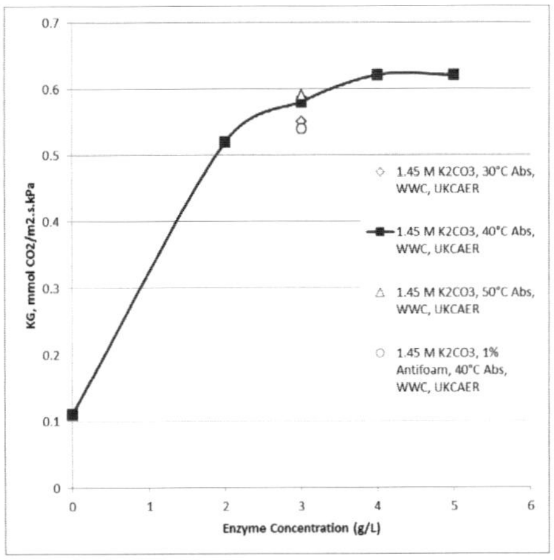

Overall Mass Transfer Coefficient (K_g) Enhanced by Enzyme in WWC

novozymes Rethink Tomorrow

Bench-scale Unit Description

- Flow Rates
  - Gas: 30 SLPM (15% CO_2)
  - Liquid : 300-600 ml/min
- Liquid Temperature
  - Absorber Inlet: 30-40°C
  - Stripper Inlet: ~65°C
  - Reboiler Oil Inlet: 90-95°C
- Stripper Pressure: 0.35 atm absolute
- K_2CO_3 Concentration: 23 wt%
- Enzyme Concentration: 0 – 4 g/L

UK Kentucky Center for Applied Energy Research

11

PFD of Integrated Bench-scale System

Capture Efficiency=

$$\frac{inlet\ CO_2\ mole\ flowrate\ -\ outlet\ CO_2\ mole\ flowrate}{inlet\ CO_2\ mole\ flowrate}$$

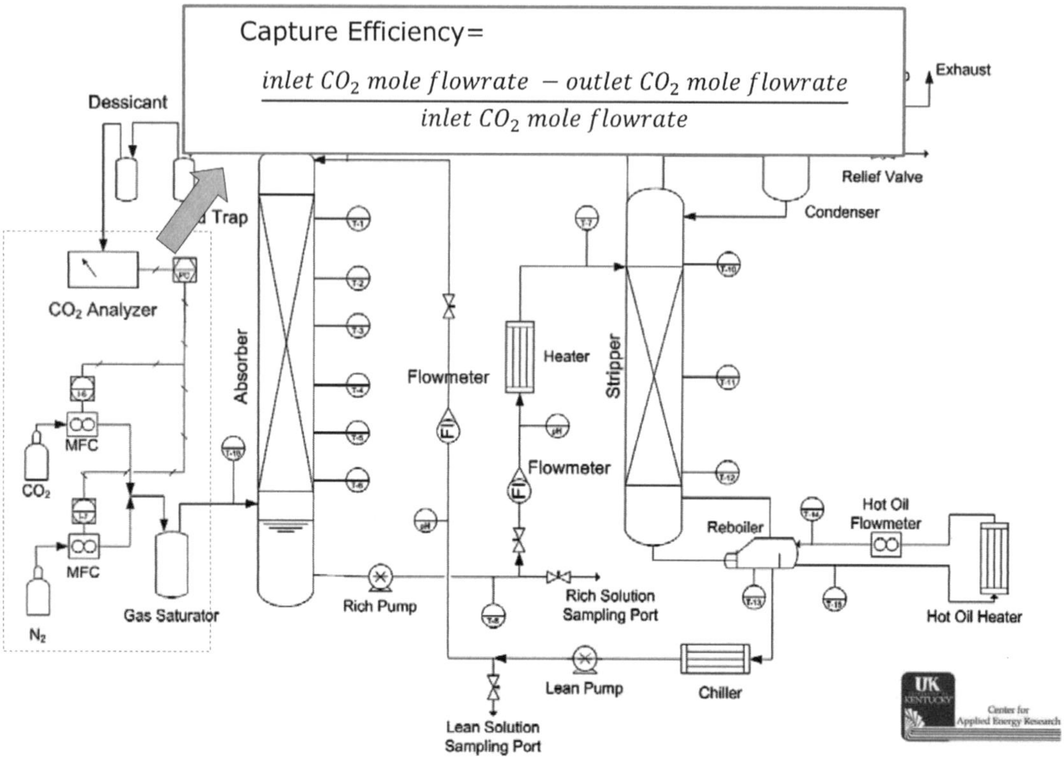

12

Bench-scale Operational Observations

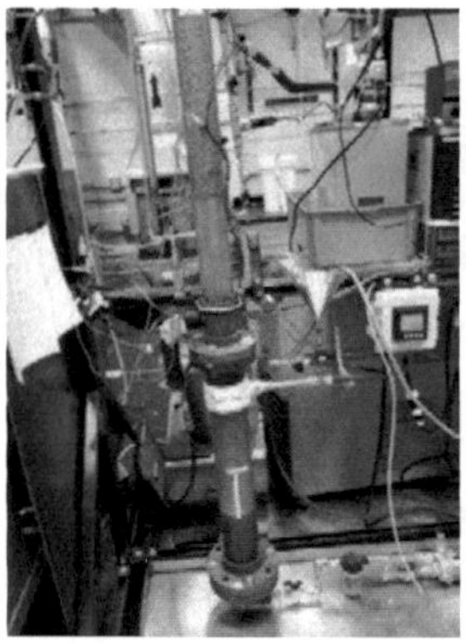

Absorber bottom

Stripper top

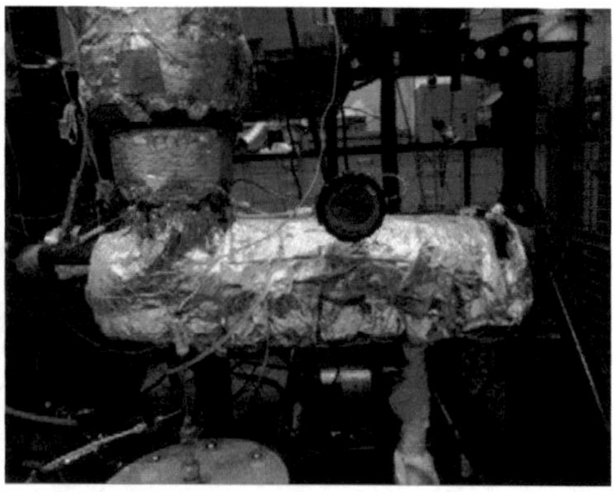

Stripper bottom

Absorber

- Low temp difference along absorber length due to low enthalpy solvent
- No visual change in packing
- Rich solvent filter removes solids

Stripper

- Water cooled condenser at top
- Tube and shell reboiler
- Antifoam dosing effectively mitigates foaming
- Bulk temp. ranges from 65°C (rich solvent inlet to stripper top) to 76°C (reboiler)
- Reboiler heating source temp. 90-95°C
- Lean solvent filter removes solids

13

novozymes
Rethink Tomorrow

Impact of Enzyme Conc. and Liquid Flow Rate

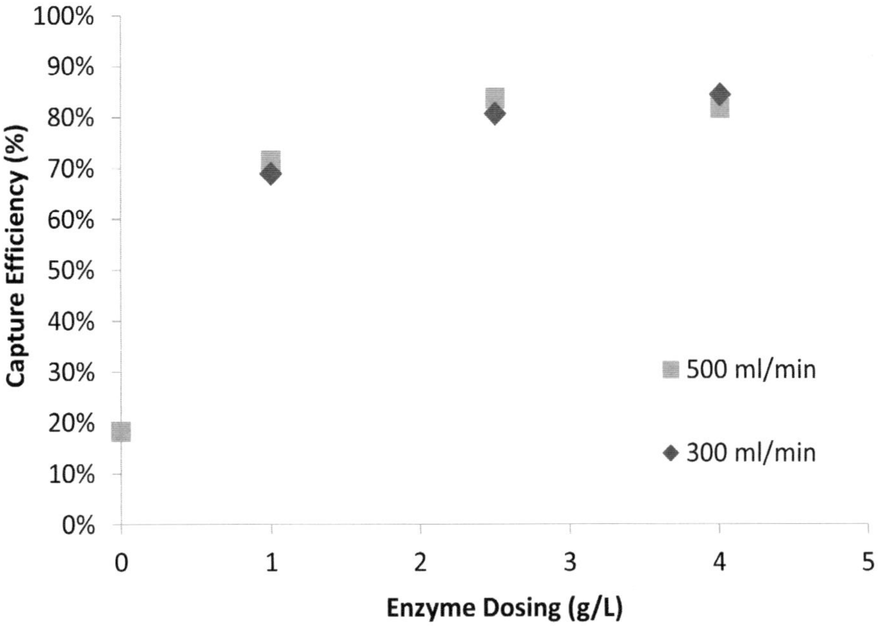

14

novozymes
Rethink Tomorrow

500 Hour Long Term Test

- Baseline conditions
  - 40°C absorber
  - 95°C reboiler heating source temperature
  - 0.35 atm absolute stripper top pressure
  - 500 ml/min liquid flow rate
  - 30 SLPM gas flow rate; 15% CO_2
  - 2.5 g/L active enzyme dose
  - 23 wt% K_2CO_3
- Solvent additions per ~7 h run day
  - Antifoam addition: ~0.04%
  - Active enzyme addition: ~20% of active enzyme dose
    - Defining enzyme replenishment is part of the program
    - Active enzyme replenishment is at least 7% of total enzyme
  - Solvent volume and alkalinity are maintained

novozymes Rethink Tomorrow

500 Hour Long Term Test Results

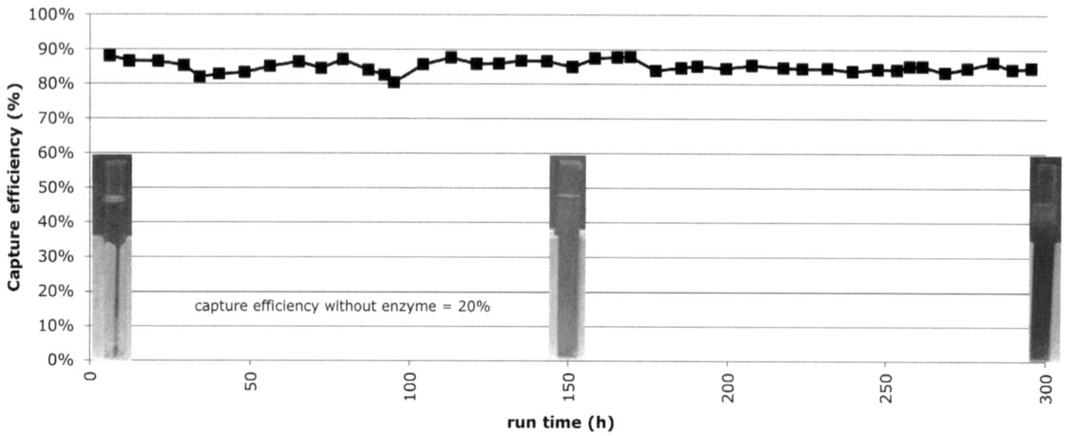

- >80% CO_2 Capture efficiency maintained
- Solvent darkens and becomes turbid
- Principle of using dissolved enzyme replenishment to achieve stable operation demonstrated

novozymes Rethink Tomorrow

Approach to Kinetic Model

- Improve existing ASPEN kinetic model for $CO_2+OH^- \rightarrow HCO_3^-$
  - Include data representing a wider temperature range than prior model
  - Include the effects of ionic strength on rate
  - Correct existing reverse kinetics to provide agreement with equilibrium model predictions at temperatures <70°C.
- Include a parallel rate expression for $CO_2+2H_2O \rightarrow H_3O^+ + HCO_3^-$
  - Model enzyme effect by accelerating this reaction, not hydroxide reaction

Comparison of equilibrium constants predicted by equilibrium model and pre-correction kinetic model.

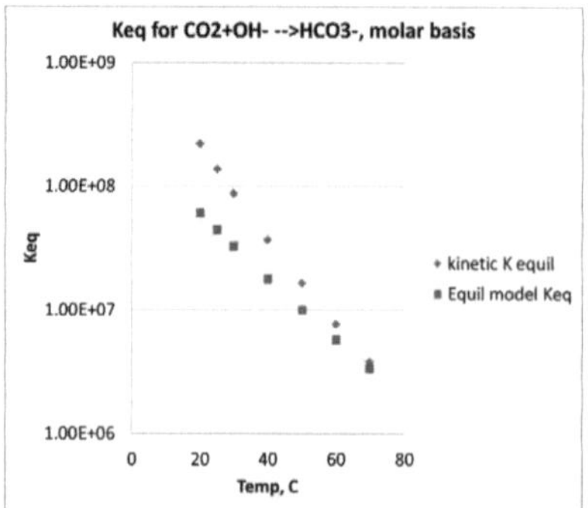

Conclusions and Next Steps

- Conclusions
  - 30 SLPM benchscale unit is operational and providing unique test data for low P/low T stripping with enzyme-enhanced K_2CO_3-based solvent
  - Parametric testing resulted in selection of 500 hour test conditions operating at >80% capture
  - Current enzyme longevity is significantly diminished by travel through stripper, but can be mitigated for test purposes by replenishment program
- Next Steps
  - Complete 500 hour testing
  - Prepare full TEA, with updated kinetics-based process simulation and ASPEN models
  - Complete EH&S assessment
- Potential Future Developments
  - Improve enzyme thermal stability
    - Immobilization or chemical modification to create physical barrier to unfolding
    - ID alternate enzyme candidates and/or protein engineering to improve T stability
    - Explore alternate process configurations to reduce enzyme loss
  - Evaluate options for increasing liquid loading capacity

14AIChE Annual Meeting, Atlanta, GA

Thank-you

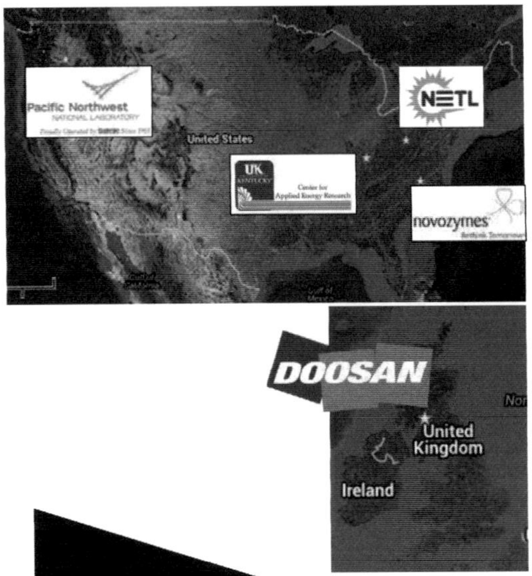

Acknowledgements

DOE-NETL
Andrew Jones

Pacific Northwest National Laboratory
Charles Freeman (PM/TL), Mark Bearden
Greg Whyatt

UK-Center for Applied Energy Research
Kunlei Liu (PM), Kun Liu (TL), Guojie Qi
Reynolds Frimpong

Doosan Power Systems
David Fitzgerald (PM), Jonathan Slater (TL)
Rafael Vidal, Tania Russell

Novozymes
Sonja Salmon (PI/PM), Alan House (TL)
Erin Yarborough

376258 Drop Size Measurements of Loaded Amine Solvents for CO2 Capture Applications

Abstract:

Spray columns can potentially be utilized for large scale CO_2 absorption in aqueous amine solvents. Drop sizes produced by commercially available nozzles are strongly dependent on the physical properties of the solvent, and they can greatly affect the rate of CO_2 absorption. Monoethanolamine (MEA) is considered as the baseline solvent for most CO_2 capture applications. The physical properties of MEA are a strong function of the loading or saturation of the solvent with CO_2. At present, there is a scarcity of drop size distribution data for loaded solutions of MEA. This work addresses this gap by evaluating the effect of CO_2 loading on the drop size distributions emerging from multiple commercially available nozzles. The radial and axial dropsize distribution measurements are made with a Phase Doppler Interferometer (PDI). The effect of MEA concentration, liquid rate, and counter-current gas rate on the drop size distribution of loaded MEA solution will be ascertained. Results will also be compared with the drop size data for unloaded MEA solutions.

Development of an Advanced Structured Packing in Absorbers for Post-Combustion CO_2 Capture

Yoshiyuki Iso *1, Jian Huang *1, Mariko Saga *1, Shinsuke Matsuno *1,
Shiko Nakamura *2, Kenji Takano *3 and Naoki Fujiwara *3

*1 Research Laboratory, IHI Corporation,
1, Shin-Nakahara-Cho, Isogo-Ku, Yokohama, Kanagawa 235-8501, Japan

*2 Products Development Center, IHI Corporation,
1, Shin-Nakahara-Cho, Isogo-Ku, Yokohama, Kanagawa 235-8501, Japan

*3 Energy & Plant Operations, IHI Corporation,
1-1, Toyosu 3-chome, Koto-ku, Tokyo 135-8710, Japan

Global warming and greenhouse gas emission reduction have become a major issue in the world. It is necessary to develop Carbon Capture and Storage (CCS) technology for coal fired power plants in order to reduce greenhouse gas emission. According to these backgrounds, Post-Combustion CO_2 Capture (PCC) technology has the great potential as the economical and efficient technology for reducing CO_2 emissions, which can be applied to both retro-fit and new construction of coal fired power plants and can control the CO_2 capture rate.

Gas-liquid interfacial flows, such as the flue gas and liquid solvent, are applied in CO_2 absorbers for PCC in order to capture CO_2 from the flue gas into liquid solvent. Efficient control of these liquid solvent flows by using packing elements in packed columns is important to increase the gas-liquid interfacial area and the mass transfer rate between the gas and the liquid. In typical packed columns, liquid is fed from the top of the packing element, and then the liquid film flow is formed on packing element surfaces. Simultaneously, gas is fed from the bottom of the packing element and makes contact counter-currently flows with the liquid film. There are two types of packing elements: structured packing and random packing. Structured packing elements can increase the gas-liquid interfacial area efficiency as well as the throughput of gas flow capacity more than random packing elements.

Control of liquid film flows by using packing elements is one of the key design factors in packed columns. In particular, the liquid channeling flow significantly reduces the gas-liquid interfacial area. To prevent this phenomenon, it is very important to predict the detailed behavior of liquid film flows for the design and development of packing elements. Therefore, the present study focuses on gas-liquid interfacial flows on an inclined plate, which is the model simplified from typical structured packing elements in absorbers for PCC.

There have previously been numerous theoretical, numerical and experimental studies on liquid film flows. These previous efforts have produced useful findings such as the liquid film shape and thickness, liquid hold-up (the amount of liquid held by a packing element), gas pressure drops, validation of Computational Fluid Dynamics (CFD) etc. However, most of these studies concerned liquid film flows on smooth surfaces, but the effects of surface texture treatments on liquid film flows have not yet been clarified. Furthermore, detailed descriptions of the transition phenomena between the film flow and the rivulet flow, as well as how such phenomena are affected by surface texture treatments, are still lacking.

In what follows, using a CFD with the Volume of Fluid (VOF) model, we investigate that the effects of surface texture treatments on the interfacial flow through the comparison of two geometry cases (smooth surface and wavy surface). Furthermore, we develop our advanced design of surface texture treatments and packing geometry. The absorption column tests using the simulated solvent (NaOH) as well as the actual amine solvent are carried out to measure the performance of the absorption rate and the gas pressure drop. Our new findings in this paper include the followings.

(1) With a typical texture geometry used in a commercial industrial packing element, the present numerical and experimental results comparing the two geometry cases (smooth surface and wavy surface) show that the surface texture treatments can help to prevent liquid channeling and can increase the wetted area quantitatively. The main reason for the increase of the wetted area on the wavy surface is that the liquid film break-up is inhibited due to the spreading of the liquid flow in spanwise direction by the wavy surface geometry.

(2) Our advanced design of surface texture treatments and packing geometry is developed. The absorption column tests using the CO_2-NaOH system show that our advanced packing trail-manufactured with two types of specified surface areas (400 m2/m3 and 200 m2/m3) have higher absorption performance and lower gas pressure drop by comparing to several conventional structured packings.

(3) We evaluate the performance of the absorption rate and the gas pressure drop of our advanced packing by using a post-combustion capture pilot plant (flue gas flow rate: 4,000 m3N/h-wet, captured CO_2 capacity: 20 tons of CO_2/day). This pilot plant can be operated in combination with the co-located coal combustion test facility, which makes it possible to carry out operational evaluations using actual flue gas and amine solvent. Our advanced packing installed in the absorber has approximately twice of the absorption performance comparing to a

conventional structured packing. And also the gas pressure drop of our advanced packing decreases approximately by half at the same absorption performance of a conventional structured packing.

On the next step, we are trying to improve not only packings but also solvents and processes, and evaluating those combined performance by using the 20 tons of CO_2/day pilot plant.

385253 Optimal CO_2 Capture Solvent Properties for Partial Capture and Natural Gas Power Plants

Abstract:

A major class of work on absorption-based capture systems is the prediction of separation plant performance for various solvents and solvent mixtures. Equilibrium or rate-based models are used to compute the imposed load (reduction in net electrical work output from a power plant) for separation using a given solvent, possibly optimizing over process parameters (such as stripper pressure). This analysis is repeated for each solvent considered. While this type of work shows the potential of each solvent, it does not show how an energetically-optimal solvent would behave.

In our modeling approach, we choose a specific process architecture, but do not choose a specific solvent. Instead of taking solvent properties as fixed, we treat them as model inputs that can be adjusted along with the system operating conditions. By adjusting the solvent properties as well as process parameters, we find the properties of the energetically-optimal solvent for a given process configuration, and the imposed load when using that solvent in that process. This value of imposed load is a more informative target than the thermodynamic minimum work or the work requirement with a specific solvent since it reveals the potential of a given process configuration.

We will discuss the properties of optimal solvents for natural gas and partial capture coal cases with varying solvent type and kinetic behavior. The model is not dependent on any particular solvent chemistry, and hence is applicable to many classes of solvents. Models for ionic liquids and non-aqueous co-solvents are obtained by setting the solvent composition appropriately. Although the model is at its core a thermodynamic model, kinetic effects can be modeled by changing the rich loading. We find that optimal thermodynamic behavior is a function of kinetic behavior and equipment size.

385038 CO_2 Capture Performance of Imidazole-Based Solvents

Abstract:

ION Engineering, The University of Alabama and other partners are currently building a slipstream-scale process for post-combustion CO_2 capture at an operating coal-fired power plant. Through a rational solvent screening process based on several key parameters, we have designed technologically and economically viable solvent compositions that can achieve significant reductions in energy penalty compared to the aqueous MEA baseline. This presentation will discuss the current status of our projects, selected properties and performance data of solvents and perspective on the market opportunities for accelerating the deployment of CO_2 capture processes.

388930 Regenerable Sorbent System for Post Combustion CO2 Capture

Abstract:

TDA Research, Inc (TDA) is developing a low cost, high capacity CO_2 adsorbent and demonstrating its technical and economic viability for post-combustion CO_2 capture for existing pulverized coal-fired power plants. TDA uses an advanced physical adsorbent to selectively remove CO_2 from the flue gas. The sorbent exhibits a much higher affinity for CO_2 than N_2, H_2O or O_2. The sorbent binds CO_2 more strongly than common adsorbents, providing the chemical potential needed to remove the CO_2. However, because CO_2 does not form a true covalent bond with the surface sites, the regeneration can be carried out with only a very small energy input (4.9 kcal per mol of CO_2) compared to amines (14.2 kcal/mol CO_2). We demonstrated the sorbent's efficacy through multiple cycles for removing CO_2 from coal derived flue gas and the sorbent achieved a stable capacity. The presence of acid gases such as SO_2 and NOx (80 and 200 ppmv, respectively) and water vapor up to 15% vol. caused no adverse effect on the CO_2 capacity. The regeneration of the sorbent and the recovery of CO_2 and its pressurization can be achieved by several approaches, including temperature and vacuum swing. A detailed techno-economic analysis using Aspen Plus^{TM} software is conducted in collaboration with University of California Irvine to estimate the impact of the TDA's CO_2 capture system on plant efficiency and cost of electricity. Our system analysis showed that TDA's CO_2 capture process will reduce plant efficiency only by 19.4% (compared to 31.2% reduction for amine scrubbers) while providing 90% capture and the increase in 1st year levelized cost of electricity (COE) is 18.4% point lower than that of amines. We will present the latest results from the DOE sponsored work at the AIChE meeting.

Performance Evaluation and Process Development of Amine Solid Sorbent for CO_2 Capture from Cement Production Plant

Jak Tanthana, Paul Mobley, Luke Coleman, and Thomas Nelson
Energy Technology Division, RTI International, Research Triangle Park, NC

Rapid increase in CO_2 concentration in the atmosphere is suggested to cause the detrimental impact to the environment and demands the CO_2 emission rate to be reduced from power and chemical industries. Cement production is known to be the second largest CO_2 emission source, following power production industry. Approximately 2/3 of the CO_2 emission from cement production is from the calcination reaction of limestone and 1/3 from fuel combustion for heating rotary kiln. Though cement flue gas shares many similarities in composition with coal-fired power plant flue gas, it contains significant higher concentration of O_2, H_2O, NO_x, and fines.

In a collaboration project with Norcem, a Norwegian cement producer, RTI was selected to develop a lab-scaled demonstration unit which utilizes the porous solid amine sorbent to evaluate the effectiveness together with long term stability of the material for CO_2 capture under cement flue gas condition. Polyethyleneimine impregnated on porous silica (PEI/SiO_2) sorbent was made in-house and subjected to multiple cycles of CO_2 absorption/regeneration using Packed-Bed Reactor system. The CO_2 capture performance and stability of PEI/SiO_2 was evaluated under simulated coal-fired flue gas, simulated cement flue gas without SO_2 and simulated cement flue gas with 100 ppm of SO_2 by performing 100 cycles of CO_2 absorption/regeneration at each flue gas condition. The result shows that PEI/SiO_2 exhibits similar CO_2 loading despite the large difference in composition between coal-fired flue gas and cement flue gas. Significant degradation and reduction in CO_2 loading was observed when 100 ppm of SO_2 was introduced to cement flue gas. Post analysis of the sorbent by ATR-IR revealed that sulfate-amine salt was formed under presence of SO_2 which reduced the amount of CO_2 absorption sites.

Development on the process aspect of this project resulted in a packed-bed reactor system with flue gas pretreatment unit equipped at the front-end of the process. The Automated Sorbent Testing Rig (ASTR) was designed specifically to handle cement flue gas with minimal involvement from user to operate. ASTR was commissioned and operated under simulated cement flue gas at RTI facility prior to installation at Norcem's cement production plant in Brevik, Norway. Once installation of flue gas feed system to ASTR is fully completed, CO_2 loading and performance stability of the sorbent will be evaluated under real cement flue gas. A set of parametric study will also be conducted on site to evaluate system performance with variation that typically found in cement flue gas quality.

Kinetics of CO_2 Capture by Novel Ionic Liquid/ Carbon Dioxide Binding Organic Liquid Hybrid Solvents

Ozge YUKSEL ORHAN, Erdogan ALPER

Hacettepe University, Chemical Engineering Department, Beytepe, Ankara, Turkey.

Because of its effect on the climate change, carbon dioxide capture and sequestration (CCS) technologies have become one of the most important global concerns. It is generally accepted that the most appropriate method that can be used today is to absorb carbon dioxide from gas mixtures into aqueous amine solutions [1]. Although, CO_2-aqueous amine process is accepted as a mature technology, it has many inherent drawbacks such as low absorption capacity, degradation, corrosion and especially high energy needs for solvent regeneration. Therefore, the developments of new solvents that can capture CO_2 efficiently, reversibly, cost-effectively and environmentally benign have gained importance.

A novel approach is to formulate blended solvents consisting of an ionic liquid and a carbon dioxide binding organic liquid 'CO_2-BOL' which can react with CO_2 reversibly. CO_2BOLs are the mixture of strong amidine or guanidine bases and n-alcohols. The primary advantages of these systems are high CO_2 loadingcapacities, low specific heat capacities and a less energy intensive solvent regeneration as compared to aqueous alkanolamine solutions. However, there is basically no information about the reaction kinetics of such systems and the aim of this presentation is to supply some of these data.

In this study, because of the most appropriate hydrodynamic properties, such as viscosity, an imidazolium based ionic liquid (1-ethyl-3-methyl imidazolium bis (trifluoromethylsulfonyl) imide ([emim][Tf_2N])) was blended with an amidine (or a guanidine)/n-alcohol to develop a novel hybrid solvent. Previously, amidine (or guanidine)/n-hexanol systems have been examined by us and it was found that they have favorable kinetics. Further details of reaction mechanism can be found in the study of Cagdas et al. [2]. It is particularly interesting that the reaction product is also an ionic liquid. In this work, as an amidine DBU (1,8-Diazabicyclo[5.4.0]undec-7-ene), DBN(1,5-Diazabicyclo[4.3.0]non-5-ene) and as a guanidine TMG (1,1,3,3-Tetramethylguanidine, TBD (1,5,7-Triazabicyclo[4.4.0]dec-5-ene) were used.

Reaction kinetics experiments were carried out by a stopped-flow apparatus (Hi-Tech Scientific, UK; Model SF-61SX2) at a temperature range of 283-303 K. The stopped flow apparatus yields directly the intrinsic rate of a rapid homogenous reaction by measuring conductivity change during the reaction. The observed pseudo-first-order-rate constant (k_o) is given by:

$$r = k_o [CO_2] \qquad (1)$$

and obtained by fitting this equation to the experimental data. To satisfy the pseudo-first-order condition, the concentration of ionic liquid and superbase were in excess of dissolved CO_2 by keeping the molar ratio about 10 to 1 for all runs. Each experimental set was repeated an average of 10 times to achieve consistent k_o values at specified conditions. A typical graphical output is shown in Figure 1 for 10.0 wt % ([emim][Tf_2N]) and 5.0 wt % TMG in 1-hexanol system at 288 K, similar outputs were obtained for other temperatures and concentration ranges studied in this work.

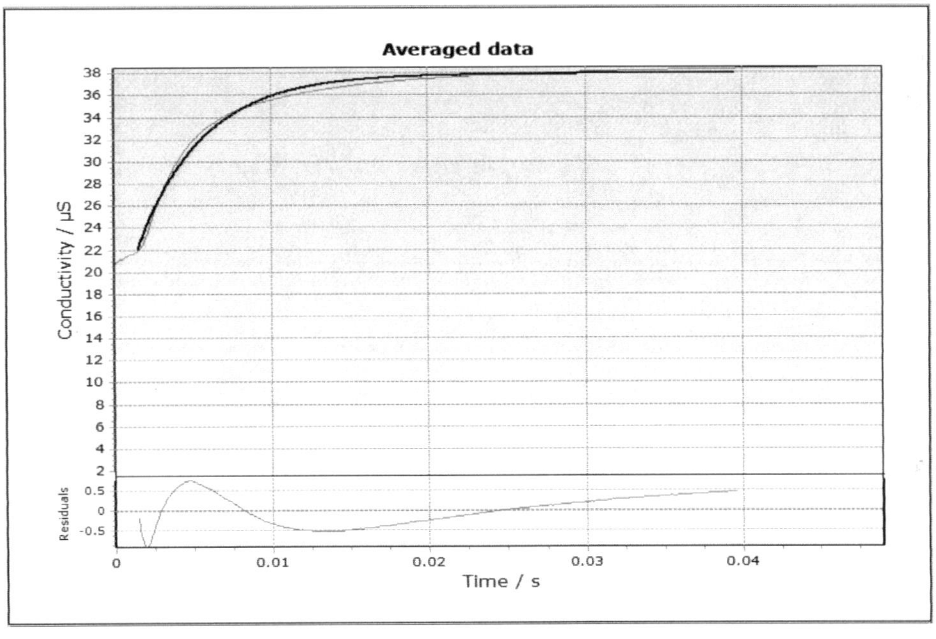

Figure 1. Typical combined graph for the 10.0 wt % ([emim][Tf_2N]) and 5.0 wt % TMG /1-hexanol system at 288K.

Experiments were carried out by keeping ionic liquid (IL) concentration constant (10.0 wt %) and by varying organic base (amidine or guanidine) percentage (5.0, 10.0,15.0 and 20.0 wt %) in a 1-hexanol medium. Based on pseudo first order reaction conditions, it was observed that the kinetic data were fitted well to the termolecular reaction mechanism. The observed reaction rate constant, k_o was found to be highest for TMG-IL/1-hexanol hybrid system. The trend of reaction rates effected by superbase type are given as (TMG> DBU> TBD> DBN). In Figure 2, the observed reaction rate constant k_o obtained at 283, 288, 298 and 303 K in 1-hexanol medium is shown as a function of TMG concentration. Increase in TMG concentration resulted in increase in the observed pseudo-first-order reaction rate constants (k_o) confirms the direct dependency of the rate constant on concentration of superbase. Also, k_o values increase progressively with an increase in the temperature for a given concentration as expected.

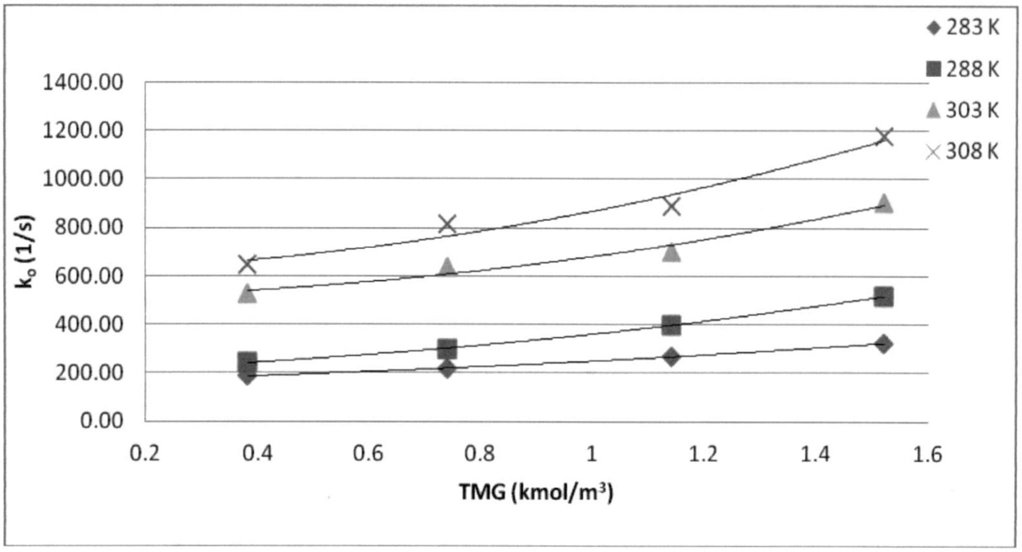

Figure 2. Effect of TMG concentration on k_o values at 298K; [IL] = 0.22 kmol/m^3

Using the least square method, empirical power law kinetics was fitted to all experimental data and fractional orders have been obtained. The orders ranging between 1.01 and 1.95 with regression value of R^2= 0.962-0.998 indicate that these results are in agreement with single-step termolecular reaction mechanism. Therefore, the experimentally obtained observed k_o values were correlated using the termolecular mechanism to determine the forward reaction rate constants k [$m^6kmol^{-2}s^{-1}$] of four different superbase (DBU, TMG, DBN, TBD) systems at specified temperatures. Furthermore, the Arrhenius diagram was plotted and the activation energies of four different superbase hybrid systems at four concentrations and four temperatures were determined. The activation energies for all systems were also calculated by evaluating the Arrhenius Equation. The average activation energies for the IL+TMG/1-hexanol, IL+TBD/1-hexanol, IL+DBU/1-hexanol and IL+DBN/1-hexanol systems were found to be 44.92 kJ/mol, 50.25 kJ/mol, 57.92 kJ/mol and 65.24 kJ/mol respectively. It was observed that the increase in organic base concentration did not have a significant effect on activation energies.

The results obtained for hybrid systems were compared with the reaction rate data of a popular amine MEA/n-hexanol. At 298 K, it was found that the rate constants k_o of 1M MEA/n-hexanol solution (810 s^{-1}) were approximately equal to corresponding concentration of IL+TMG/1-hexanol hybrid solution. The result of this study shows that these nonaqueous novel hybrid systems are potential candidates as alternative solvents to conventional carbon dioxide capture agents.

References

[1] Tontiwachwuthikul, P.; Idem, R. Recent Progress and New Developments in Post-Combustion Carbon-Capture Technology with Reactive Solvents, Future Science, 239 pages, October, 2013.

[2] Ozturk, M.C.; Yuksel Orhan, O.; Alper, E.Int. J. Greenh. Gas Con.,In Press, 2014 (http://dx.doi.org/10.1016/j.ijggc.2014.04.023).

CO_2 CAPTURE FROM FOSSIL FUEL POWER PLANT FLUE GAS USING AQUEOUS SOLUTION OF RENEWABLY PREPARED ALKANOLAMINE

Rahul Bhosale[1,], Vijaykumar Mahajani[2], Anand Kumar[1], LJP van den Broeke[1], Shahd Gharbia[1], Mehak Jilani[1], Jamila Folady[1], Dareen Dardor[1], Diana Abuarja[1], Aeman AlJumily[1], Amira Nemmour[1], Dalia Abo Mazid[1], Narimane Belloula[1], Aya Said[1]*

[1]Department of Chemical Engineering, College of Engineering, Qatar University, PO Box 2713, Doha, Qatar.
[2]Department of Chemical Engineering, Institute of Chemical Technology, Mumbai, India.
**Corresponding Author: rahul.bhosale@qu.edu.qa*

Abstract

In this study, $K_2CO_3/KHCO_3$ as hydroxyl radical scavengers was mixed with aqueous EAE to form a new absorbing solvent i.e. EHRS solvent which will be used for the reactive absorption of CO_2. Degradation of the EHRS solvent in presence of O_2 (oxidative degradation) was examined by employing accelerated experimental conditions at 150°C. The physico-chemical properties such as density, viscosity, and diffusivity and solubility of CO_2 in EHRS solvent at various experimental conditions were estimated. Furthermore, kinetics of reactive absorption of CO_2 in EHRS solvent was explored by using a stirred cell reactor using a fall in CO_2 pressure technique.

Introduction

Atmospheric concentration of CO_2 (a major greenhouse gas) has reached up to 392 ppmv as of 2011, which was 280 ppmv in pre-industrial era and if this trend continues, by 2050, CO_2 concentration is expected to increase upto 550 ppmv [1-2]. Among many, flue gas from power plants, cement manufacturing, refineries, and etc. is considered as one of the major sources of CO_2 emissions. Excessive utilization of fossil fuels contribute to 64% of greenhouse gas emission worldwide and fossil fuel based power generation currently accounts for over one third of global annual CO_2 emissions. The continuous emission of CO_2 is considered as one of the significant reasons for the greenhouse effect which leads to global warming, glaciers melting, sea level rise, and extreme climates.

In carbon sequestration science, the separation and capture of CO_2 is considered as one of the highest priorities as the capture cost is expected to be up to 75% of the total costs for geological and oceanic sequestration. Reactive absorption of CO_2 from process gas streams and flue gas from various sources by using aqueous alkanolamines is considered as the most mature post-combustion CO_2 capture technique with maximum efficiency and lower complexity [1-11]. A typical alkanolamine based reactive absorption/desorption process useful for the removal of CO_2 from power plant flue gas is shown in Figure 1. As an alternative for sequestration, captured CO_2 can be reutilized via solar thermochemical cycles to produce syngas which can be further processed to liquid fuels such as Methanol, Jet Fuel via the Fischer-Tropsch process [12-30].

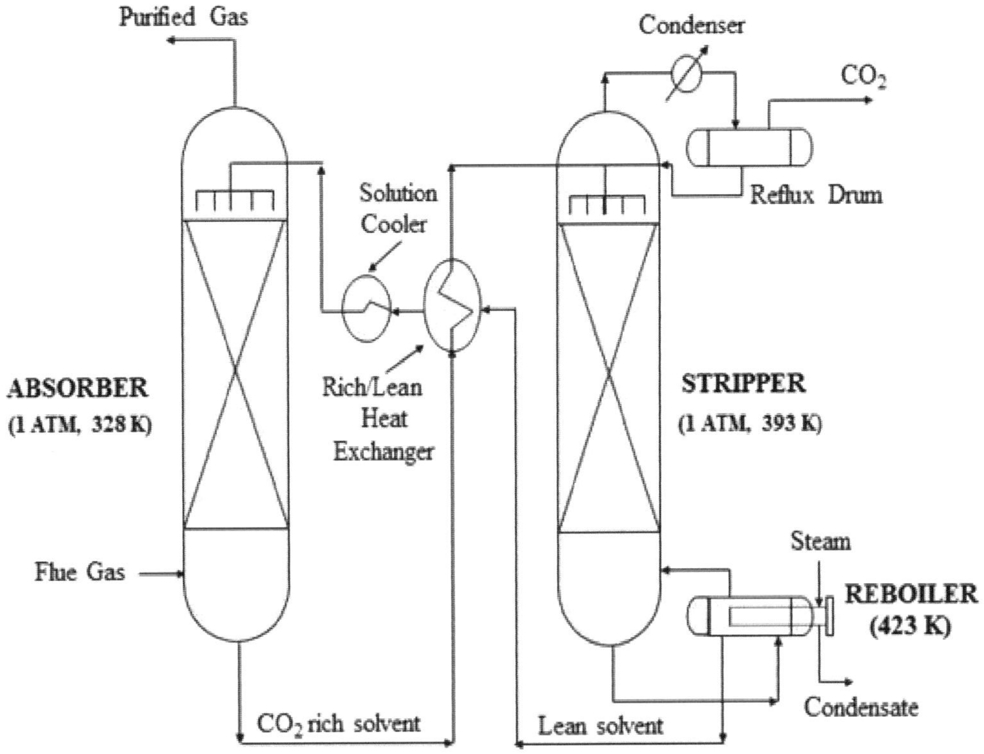

Fig. 1. A typical alkanolamine based reactive absorption/desorption process useful for the removal of CO_2 from power plant flue gas [2].

The reactive absorption/desorption process shown in Fig. 1 is a two-step process where the first step belongs to the absorption of CO_2 by a suitable absorbent at 40 – 70^oC and the second step corresponds to the desorption/stripping of CO_2 from the absorbing solvent by using thermal energy (temperatures in the range of 100 – 150^oC). There are a number of solvents, which are being used for CO_2 absorption and among which conventionally aqueous alkanolamines are the best chemical solvents. The hydroxyl group of alkanolamine serves to reduce the vapour pressure and increase water solubility, while the amino group provides the necessary alkalinity in water solution for the absorption of acidic CO_2. Aqueous solution is useful because reactions of CO_2 and chemical solvents are ionic in nature and chemical solvents i.e. polar solvents are soluble in polar solvents.

Several alkanolamines such as monoethanolamine (MEA), diethanolamine (DEA), di-isopropanolamine, (DIPA), methyldiethanolamine (MDEA), and diglycolamine (DGA) have been used industrially for a number of years. Among all previously studied alkanolamines, MEA remains the most popular for CO_2 capture owing to its higher reactivity and lower solvent cost as compared with other alkanolamines. However, MEA also possesses several limitations such high vapor pressure which makes it unsuitable for low pressure operations, inferior CO_2 loading capacity, higher oxidative degradation and solvent regeneration cost, corrosion in equipments and pipelines, and etc. Due to all such shortcomings, quest for an alternative for MEA is inevitable.

In our recent investigations, we observed that ethylaminoethanol (EAE) is a promising alternative for MEA due to its lower solvent decomposition during thermal degradation in presence of CO_2, higher absorption capacity and suitable reaction kinetics. The major raw material for the manufacturing of EAE is ethanol, which in turn, can be prepared from agricultural products and/or residues; hence EAE represents a promising absorbent as it can be prepared from renewable resources [1-2]. Due to its lower molecular weight as compared to MEA, EAE is less volatile and subsequently vaporization losses will be lower. Furthermore, the heat of reaction of EAE-CO_2 reactive system is lower, as compared to the MEA-CO_2 reactive system, which indicates that the less heat is required for CO_2 stripping which further reflects into lower solvent regeneration cost.

In this paper, we report the preparation of a formulated solvent containing EAE (chemical solvent) and K_2CO_3/$KHCO_3$ (hydroxyl radical scavenger, useful to decrease the oxidative degradation) for capture of CO_2 from various process streams. This formulated solvent is hereafter termed as EHRS solvent (E = ethylaminoethanol, HRS = hydroxyl radical scavenger). Oxidative degradation of EHRS solvent was studied at 150°C (maximum temperature used in a stripper). Furthermore, the kinetics of absorption of CO_2 in EHRS solvent was also investigated.

Experimental

Materials

EAE (99% pure), anhydrous K_2CO_3 (analytical reagent grade) and $KHCO_3$ (laboratory grade) were procured. Nitrous oxide (N_2O), N_2, O_2, and CO_2 with minimum stated purities equal to 99.5% were obtained and used during degradation and kinetics experiments. All the chemicals were used without any pretreatment and all the solutions were prepared using distilled water.

Oxidative degradation experimental set-up and procedure

A high-temperature high-pressure subcritical reactor (Parr Instrument Company, Moline, USA) made up of SS 316 with 600 ml capacity was used to study the oxidative degradation of EHRS solvent. To impersonate the typical industrial flue gas-treating operating conditions, high O_2 pressures were used in this investigation. The temperature accuracy of the controller was within (± 0.5%). For a typical oxidative degradation experiment, 300 ml of EHRS solvent was loaded in the reactor. The charged reactor was first flushed with inert N_2 to ensure that the initial O_2 present in the reactor was removed and the reactor conditions were inert. After achieving the inert atmosphere, the gas inlet and outlet lines were closed and the EHRS solvent was simultaneously stirred at the predetermined speed of agitation (to eliminate the diffusion and mass-transfer limitations) and heated up to 150°C (desired degradation reaction temperature). O_2 was then sparged into the liquid phase directly beneath the impeller at a desired partial pressure. Due to the oxidative degradation of EHRS solvent, O_2 pressure declined. Hence, to maintain the desired O_2 pressure throughout the duration of reaction, the reaction chamber was boosted with an additional O_2 supply. To estimate the % degradation of EHRS solvent, about 4 ml of the reaction mixture sample was withdrawn from the reactor periodically and analyzed using gas chromatography (GC).

Solubility and kinetics experimental set-up and procedure

Kinetics of the absorption of CO_2 in EHRS solvent was investigated by performing experiments using a stirred cell reactor as shown in Fig. 2. Absorption experiments were carried out by operating the reactor in a batch mode with flat horizontal gas-liquid interface. A novel fall in CO_2 pressure method was employed to investigate the absorption kinetics and for this purpose a pressure transducer was used (mounted on the flange and coupled with a data acquisition system). For the investigation of kinetics of absorption of CO_2, the reactor was first charged with 250 ml of EHRS solvent and the reactor contents were allowed to attain the desired absorption temperature. As a starting point of the reaction, CO_2 with a preferred partial pressure was loaded inside the reactor and the initial pressure of the reaction chamber was recorded by the data acquisition system. The reactor content was then stirred at a pre-determined speed of agitation while ensuring that the gas-liquid interface was not disturbed. Fall in the system pressure due to the reactive absorption of CO_2 in EHRS solvent was monitored and recorded with the help of pressure transducer and data acquisition system. This measurement, which is thoroughly based on the fall in pressure of CO_2, enables a simple and straightforward estimation of the absorption rate. Further analysis of the liquid phase is not required and the decrease in the system pressure is the only factor considered for the evaluation of kinetic parameters. Similar to the absorption experiments, the solubility of CO_2 in EHRS solvent was estimated also by using the same stirred cell reactor and by employing N_2O analogy [31]. The reproducibility of the results during solubility and kinetic experiments was examined and the error for all measurements was observed to be less than ± 2%.

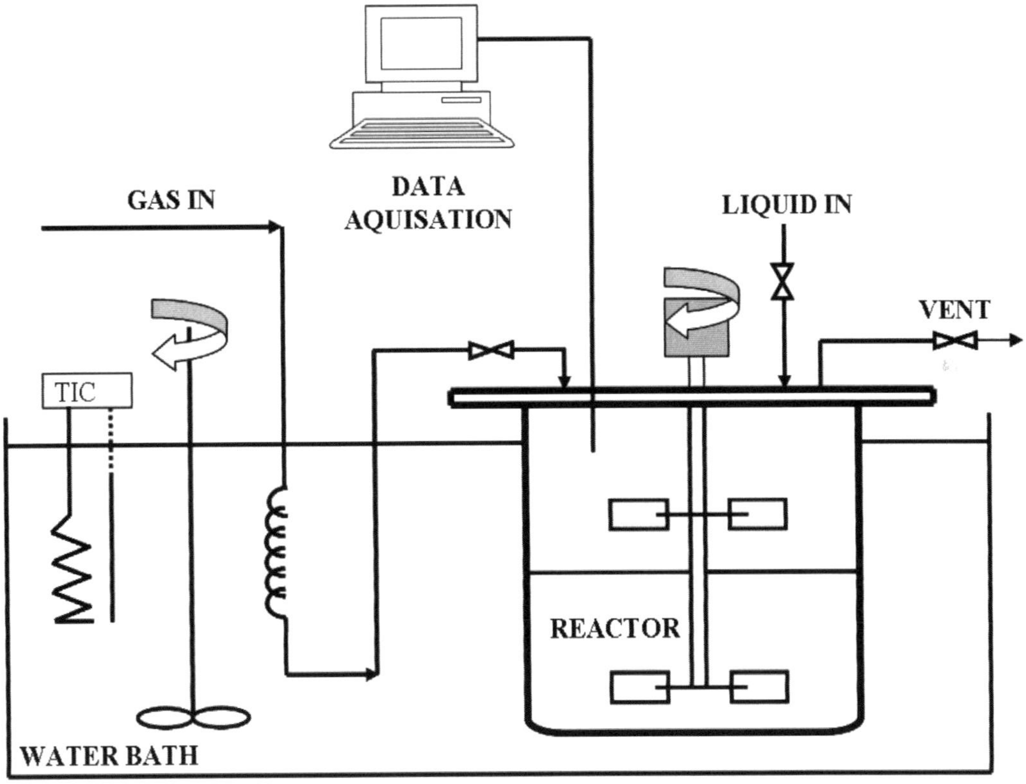

Fig. 2. Experimental set-up for solubility and kinetics study.

Results and Discussion

Oxidative degradation

Accelerated oxidative degradation of EHRS solvent was studied and compared with the degradation of aqueous EAE (without hydroxyl radical scavengers) at identical operating conditions reported in Table 1. The liquid sampled obtained during the oxidative degradation experiments were analyzed using gas chromatography technique (GC). For this purpose a GC (Chemito 8610, Toshniwal Instruments) equipped with a Flame Ionization Detector (FID) was used. A Tenax GC column (length = 1.3 m and inner diameter = 3.8 mm) was used during GC analysis for the separation of components.

Table 1. A comparative study of oxidative degradation of aqueous MEA, aqueous EAE, and EHRS solvent: Experimental parameters.

Experimental parameters	Degradation of aqueous EAE	Degradation of EHRS solvent
Amine concentration	5000 ppm	5000 ppm
$K_2CO_3/KHCO_3$	None	10wt% of EAE
Temperature	150°C	150°C
Partial pressure of O_2	6.8 bar	6.8 bar
Total reactor pressure	12.3 bar	12.3 bar
Agitation speed	1200 rpm	1200 rpm
Reaction time	16 h	16 h

Results obtained after analyzing the liquid samples collected from the high temperature high pressure reactor using GC are presented in Fig. 3. From the presented results, it was observed that aqueous EAE undergoes 65% and 100% degradation after 8 and 16 hours of operation. In contrast to this, EHRS solvent undergoes 24% and 39% degradation after 8 and 16 hours of operation. As the % degradation of EHRS solvent is significantly low as compared to aqueous EAE, it confirms that the addition of $K_2CO_3/KHCO_3$ as hydroxyl radical scavengers substantially suppresses the oxidative degradation of both EAE and NMP.

Physico-chemical properties

It is highly essential to know the physico-chemical properties of a solvent such as density, viscosity, and solubility and diffusivity of CO_2 for the investigation of kinetic parameters for the reactive absorption of CO_2. Hence, the physico-chemical properties of the EHRS solvent were determined at various concentrations and temperatures. Density and viscosity of the EHRS solvent were measured using density meter and an Ostwald's capillary viscometer. The diffusivity and solubility of CO_2 in EHRS solvent was estimated by using N_2O analogy. According to N_2O analogy, the ratios of the CO_2/N_2O diffusivities and CO_2/N_2O solubilities in water are equal to those in the amine solutions [31]. Furthermore, the liquid side mass transfer coefficient for the CO_2 – EHRS reaction system was determined using the techniques previously used by our group and explained elsewhere [1, 9].

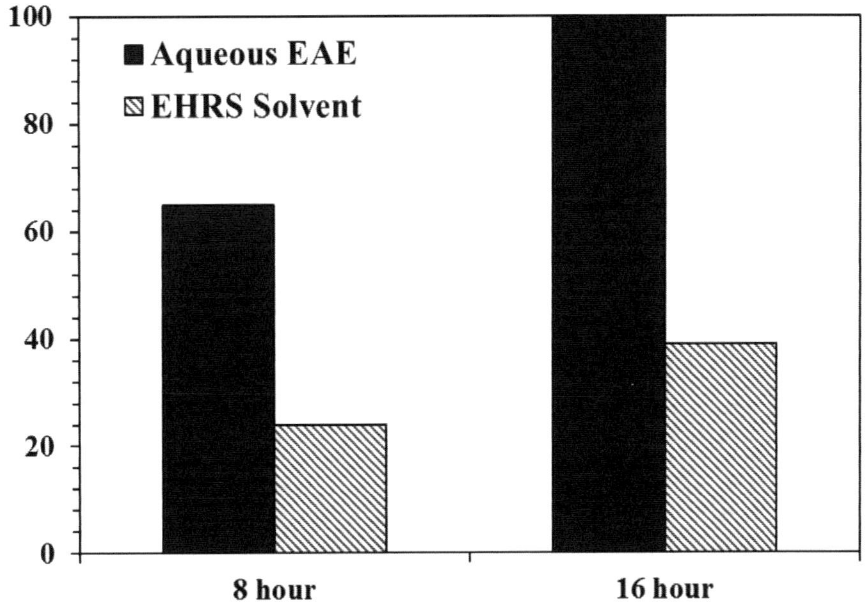

Fig. 3. Comparison between aqueous EAE and EHRS solvent in terms of oxidative degradation.

Absorption kinetics

Although the addition of $K_2CO_3/KHCO_3$ as hydroxyl radical scavengers has improved the performance of EHRS solvent by reducing the oxidative degradation significantly, it is very important to understand the reaction kinetics and CO_2 absorption capacity of this solvent. Therefore, the kinetics of reactive absorption of CO_2 in EHRS solvent was researched by using a stirred cell reactor and fall in CO_2 pressure technique. To compare the CO_2 absorption kinetics with aqueous EAE, the experimental conditions were kept similar to our previous investigation [1]. Experimental conditions employed are listed in Table 2.

Table 2. Experimental parameters employed during the study of kinetics of reactive absorption of CO_2 in EHRS solvent.

Experimental parameters	**Kinetics of CO_2-EHRS reaction system**
EAE concentration	0.5 to 2 kmol/m^3
$K_2CO_3/KHCO_3$	10wt% of EAE
Temperature	30°C
Partial pressure of CO_2	4 kPa
Agitation speed	80 rpm

At first, the effect of stirring speed on rate of absorption of CO_2 in EHRS solvent was probed and the experimental outcomes shows no variation in the absorption rate due to the change in the stirring speed in the range of 60 to 120 rpm. This observation confirms that the CO_2-EHRS reaction system belongs to the fast reaction regime. For further investigation, all the absorption experiments were carried out at stirring speed of 80 rpm.

Kinetic parameters for the CO_2-EHRS reaction system were estimated by examining the effect of EAE concentration on the CO_2 absorption rate. From the experimental, It was realized that as the concentration of EAE increased from 0.5 to 2 kmol/m^3, the rate of absorption of CO_2 was also enhanced by 40.36%. However, the reaction rate constant was observed to be constant at 8783 m^3/kmol·sec irrespective of the increase of the initial EAE concentration. When compared with rate of absorption of CO_2 in aqueous EAE [1], the rate of absorption of CO_2 in EHRS solvent seems identical and a very negligible difference was observed.

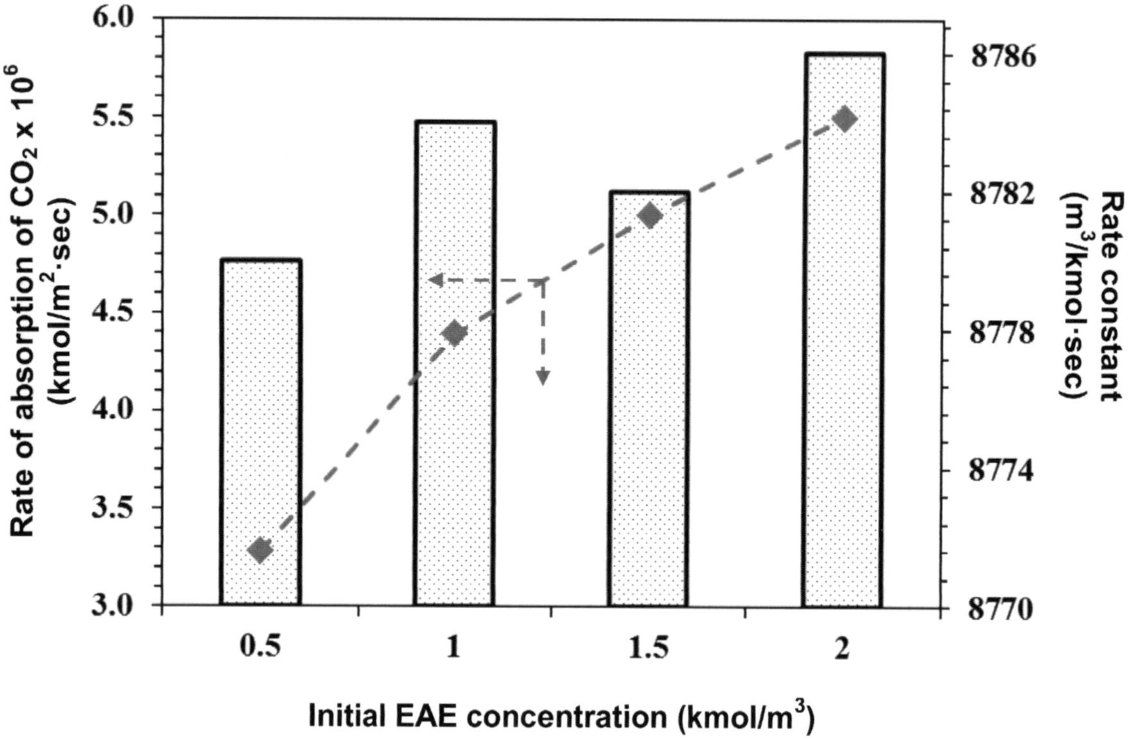

Fig. 4. Effect of initial EAE concentration on rate of absorption of CO_2 and rate constant.

Conclusions

Effect of addition of $K_2CO_3/KHCO_3$ as hydroxyl radical scavengers on oxidative degradation of EHRS solvent was investigated at 150°C. From obtained results it was understood that the % oxidative degradation of EHRS solvent (39%) was significantly low as compared to the aqueous EAE solvent (100%) after 16 hours of reaction. Furthermore, the kinetics of absorption of CO_2 in EHRS solvent was also investigated. Effect of concentration of EAE on absorption rate and rate constant was studied and it was observed that the rate of absorption of CO_2 increases with the increase in the EAE concentration, however the rate constant value remains unaffected. At the end it can be concluded that due to the addition of $K_2CO_3/KHCO_3$ as hydroxyl radical scavengers, the oxidative degradation of EHRS was reduced significantly without affecting the reaction kinetics for the CO_2-EHRS reaction system.

Acknowledgments

Authors wish to acknowledge the University Grants Commission (Government of India), Institute of Chemical Technology (ICT, Mumbai), and Qatar University for the support provided during the course of this study.

References

1. Bhosale, R., and Mahajani, V. (2013), "Kinetics of absorption of carbon dioxide in aqueous solution of ethylaminoethanol modified with N-methyl-2-pyrolidone", *Separation Science & Technology*, **48**, 2324.
2. Bhosale, R., and Mahajani, V. (2013), "Kinetics of thermal degradation of renewably prepared amines useful for flue gas treatment", *Journal of Renewable and Sustainable Energy*, **5**, 063110-1.
3. Gouedard, C., Picq, D., Launay, F., and Carrette, P. (2012), "Amine degradation in CO_2 capture. I. A review", *International Journal of Greenhouse Gas Control*, **10**, 244.
4. Vaidya, P., and Kenig. E. (2007), "CO_2-alkanolamine reaction kinetics: A review of recent studies", *Chemical Engineering & Technology*, **30**, 1467.
5. Li, J., Henni, A., and Tontiwachwuthikul, P. (2007), "Reaction kinetics of CO_2 in aqueous ethylenediamine, ethyl ethanolamine, and diethyl monoethanolamine solutions in the temperature range of 298−313 K, using the stopped-flow technique", *Industrial & Engineering Chemistry Research,* **46**, 4426.
6. Mimura, T., Suda, T., Iwaki, I., Honda, A., and Kumazawa, H. (1998), "Kinetics of reaction between carbon dioxide and sterically hindered amines for carbon dioxide recovery from power plant flue gases", *Chemical Engineering Communication*, **170**, 245.
7. Bavbek, O., and Alper, E. (1999), "Reaction mechanism and kinetics of aqueous solutions of primary and secondary alkanolamines and carbon dioxide," *Turkish Journal of Chemistry*, **23**, 293.
8. Vaidya, P., and Kenig. E. (2007), "Absorption of CO_2 into aqueous blends of alkanolamines prepared from renewable resources," *Chemical Engineering Science*, **62**, 7344.
9. Vaidya, P., and Mahajani, V. (2005), "Kinetics of the reaction of CO_2 with aqueous formulated solution containing monoethanolamine, N-methyl-2-pyrrolidone, and diethylene glycol", *Industrial & Engineering Chemistry Research,* **44**, 1868.
10. Sutar, P., Jha, A., Vaidya, P., and Kenig. E. (2012), "Secondary amines for CO_2 capture: A kinetic investigation using N-ethylaminoethanol", *Chemical Engineering Journal*, **207-208**, 718.
11. Bhosale, R., Gharbia, S., Folady, J., Jilani, Mehak., Dardor, D., Kumar, A., van den Broeke, L. J. P., Mahajani, V., and Dash, S. K. (2014), "CO_2 capture using an aqueous formulated solvent containing ethylaminoethanol, N-Methyl-2-Pyrolidone, and hydroxyl radical scavengers: study of solvent degradation and absorption kinetics", *Proceedings of 4th Annual Gas Processing Symposium,* (Accepted – In Press).
12. Loutzenhiser, P., and Steinfeld, A. (2011), "Solar syngas production from CO_2 and H_2O in a two-step thermochemical cycle via Zn/ZnO redox reactions: Thermodynamic cycle analysis", *International Journal of Hydrogen Energy*, **36**, 12141.

13. Stamatiou, A., Loutzenhiser, P., and Steinfeld, A. (2010), "Solar syngas production from H_2O and CO_2 via two-step thermochemical cycles based on Zn/ZnO and FeO/Fe_3O4 redox reactions: Kinetic analysis", *Energy & Fuels*, **24**, 2716.
14. Loutzenhiser, P., Meier, A., and Steinfeld, A. (2010), "Review of the two-step H_2O/CO_2-splitting solar thermochemical cycle based on Zn/ZnO redox reactions", *Materials*, **3**, 4922.
15. Abanades, F. (2012), "CO_2 and H_2O reduction by solar thermochemical looping using SnO_2/SnO redox reactions: Thermogravimetric analysis", *International Journal of Hydrogen Energy*, **37**, 8223.
16. Abanades, F., and Villafan-Vidales, H. (2011), "CO_2 and H_2O conversion to solar fuels via two-step solar thermochemical looping using iron oxide redox pair", *Chemical Engineering Journal*, 175, 368.
17. Scheffe, J., Li, J., and Weimer, A. (2010), "A spinel ferrite/hercynite water-splitting redox cycle", *International Journal of Hydrogen Energy*, **35**, 3333.
18. Bhosale, R., Shende, R., and Puszynski, J. (2010), "H_2 generation from thermochemical water splitting using sol-gel derived Ni-ferrite", *Journal of Energy and Power Engineering*, **4**, 27.
19. Bhosale, R., Shende, R., and Puszynski, J. (2010), "H_2 generation from thermochemical water-splitting using sol-gel synthesized Zn/Sn/Mn-doped Ni-ferrite", *International Review of Chemical Engineering*, **2**, 852.
20. Bhosale, R., Shende, R., and Puszynski, J. (2012), "Thermochemical water-splitting for H_2 generation using sol-gel derived Mn-ferrite in a packed bed reactor", *International Journal of Hydrogen Energy*, **37**, 8223.
21. Bhosale, R., Khadka, R., Shende, R., and Puszynski, J. (2011), "H_2 generation from two-step thermochemical water-splitting reaction using sol-gel derived $Sn_xFe_yO_z$", *Journal of Renewable and Sustainable Energy*, **3**, 063104-1.
22. Bhosale, R., Shende, R., and Puszynski, J. (2012), "Sol-gel derived $NiFe_2O_4$ modified with ZrO_2 for hydrogen generation from solar thermochemical water-splitting reaction", *Proceedings of Materials Research Society Symposium*, 1387.
23. Bhosale, R., Alxneit, I, van den Broeke, Leo L. P., Kumar, A., Jilani, M., Gharbia, S., Folady, J., and Dardor, D. (2014), "Sol-gel synthesis of nanocrystalline Ni-Ferrite and Co-Ferrite redox materials for thermochemical production of solar fuels", *Proceedings of Materials Research Society Symposium* (Accepted – In Press).
24. Fresno, F., Yoshida, T., Gokon, N., Fernandez-Saavedra, R., and Kodama, T. (2010), "Comparative study of the activity of nickel ferrites for solar hydrogen production by two-step thermochemical cycles", *International Journal of Hydrogen Energy*, **35**, 8503.
25. Abanades, S., and Flamant, G. (2006), "Thermochemical hydrogen production from a two-step solar-driven water-splitting cycle based on cerium oxides", *Solar Energy*, **80**, 1611.
26. Scheffe, J., Jacot, R., Patzke, G., and Steinfeld, A. (2013), "Synthesis, characterization and thermochemical redox performance of Hf, Zr and Sc doped ceria for splitting CO_2", *Journal of Physical Chemistry C*, **177**, 24104.
27. Chueh, W. C., Falter, C., Abbott, M., Scipio, D., Furler, P., Haile, S., and Steinfeld, A. (2010), "High-flux solar-driven thermochemical dissociation of CO_2 and H_2O using nonstoichiometric ceria", *Science*, **330**, 1797.

28. Le Gal, A., and Abanades, S. (2011), "Catalytic investigation of ceria-zirconia solid solutions for solar hydrogen production", *International Journal of Hydrogen Energy*, **36**, 4739.
29. Abanades, S., and Gal, A. L. (2012), "CO_2 splitting by thermo-chemical looping based on $Zr_xCe_{1-x}O_2$ oxygen carriers for synthetic fuel generation", *Fuel*, **102**, 180.
30. Le Gal, A., Abanades, S., and Flamant, G. (2011), "CO_2 and H_2O splitting for thermochemical production of solar fuels using nonstoichiometric ceria and ceria/zirconia solid solutions", *Energy & Fuels*, **25**, 4836.
31. Versteeg, G.F., and van Swaaji, W.P.M. (1988), "Solubility and diffusivity of acid gases (carbon dioxide, nitrous oxide) in aqueous alkanolamine", *Journal of Chemical Engineering Data*, **33**, 29.

Oxy-Combustion Coal Based Power Plants: Study of Operating Pressure, Oxygen Purity and Downstream Purification Parameters

Rengarajan Soundararajan*[a], Truls Gundersen[a], Mario Ditaranto[b]

[a]Department of Energy and Process Engineering, Norwegian University of Science and Technology, Trondheim, Norway.

[b]SINTEF Energy Research, N-7465; Trondheim, Norway

*rengarajan.soundararajan@ntnu.no

Oxy-combustion coal based power plants are considered one of the cost effective ways to produce CO_2 emission free (captures at least 90 % of the CO_2 produced) electricity. Recovery of compression heat and the latent heat from the flue gases can further improve the efficiency of such power plants. Power plants that employ a boiler operating at a pressure higher than ambient pressure avoid air leakage into the boiler and also recover more of the latent heat available in the flue gases. Such power plants require compression of the oxygen stream before the combustor, an entirely new boiler house design and modifications to the downstream flue gas processing systems. Selection of an operating pressure that is energy optimal is the key to design of such systems. Besides the operating pressure, the oxygen purity and the downstream separation parameters have a considerable impact on the overall performance of such power plants. In this paper, the impact of operating pressure, oxygen purity and downstream purification parameters on the performance of oxy-combustion coal based power plants are studied.

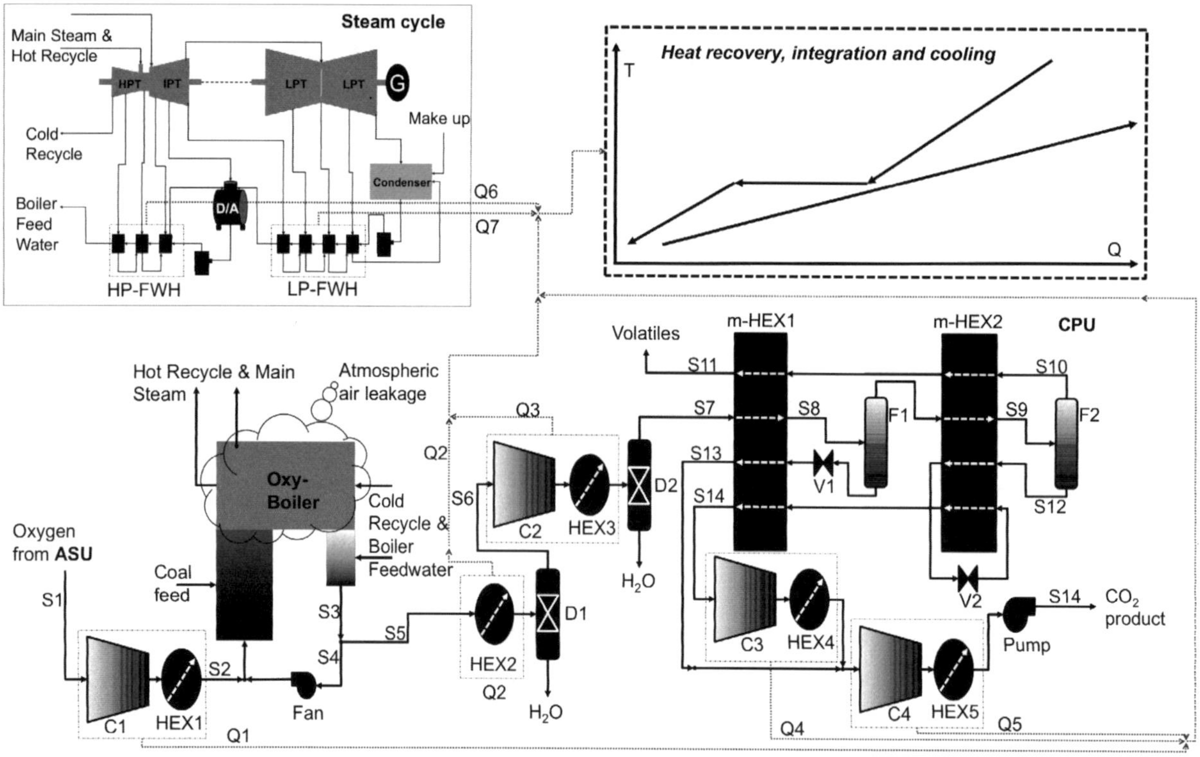

Scope

The main focus of this work is to identify near optimal values in terms of thermal efficiency for the boiler operating pressure (pressure at S2 in the figure), the oxygen purity (at S1), and CPU operating parameters such as the pre-compression pressure (pressure at S7 in the figure) and the temperature at S8. Issues related to the investment cost and operating economics are beyond the scope of this work.

Results

Adiabatic compression of the oxygen stream from the ASU (S1) to 16 bars is found to be near optimal for the overall performance of the system. Alternatively, isothermal compression using multiple stages and intercooling can be used to achieve similar efficiency values. In case of isothermal compression, preheating of the compressed oxygen is required to achieve the best efficiency. Owing to the simplicity, adiabatic compression of the oxygen is recommended as the best option. The temperature at S8 is found to have very minimal impact on the overall performance of the system. The impact of oxygen purity and pre-compression pressure is shown in the figures below.

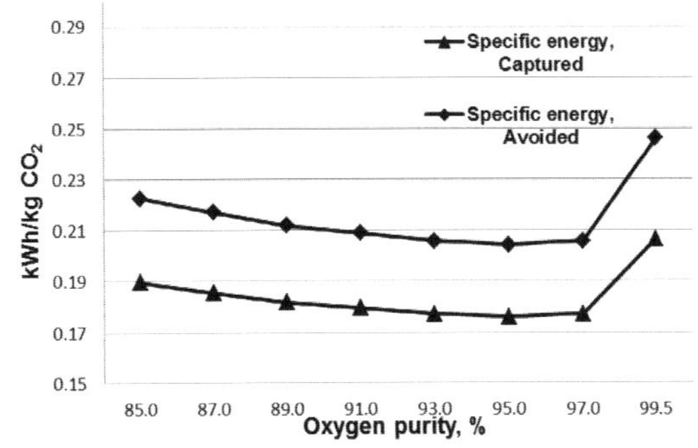

(a)

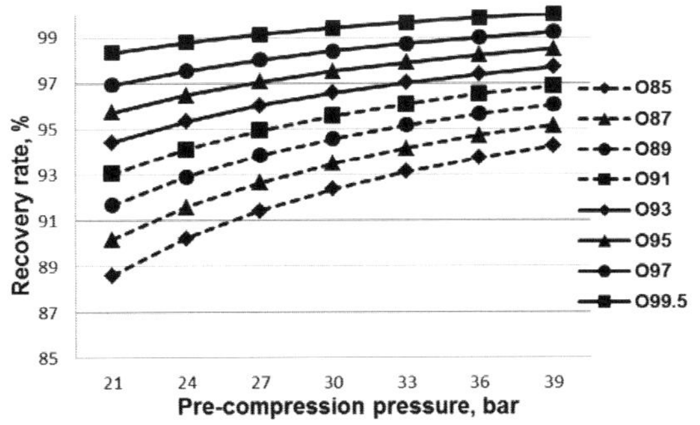

(b)

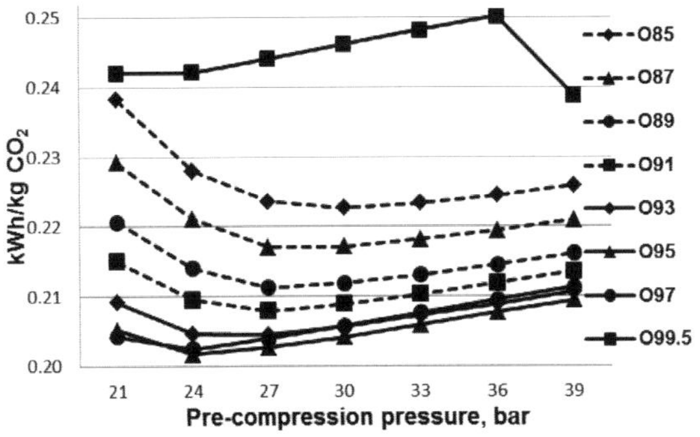

(c)

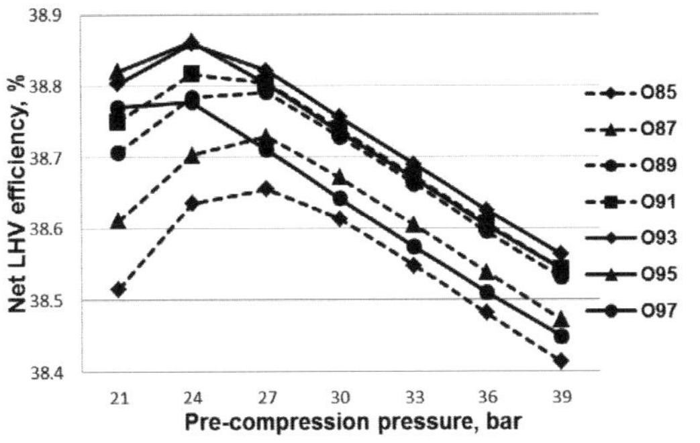

(d)

Conclusion

The baseline power plant without CO_2 capture has a net LHV efficiency of 44.7 % and the baseline oxy-combustion power plant with capture has a net LHV efficiency of 37.9 %. The low energy penalty (6.8 % points) of the capture plant is due to flue gas heat recovery and CPU heat integration. Increasing the boiler operating pressure to 16 bars recovers the flue gas latent heat and the compression heat thereby resulting in a net LHV efficiency of 38.7 %. A study of the impact of oxygen purity and CPU operating parameters provides a near optimal combination of these parameters (~97 % oxygen purity and 24 bars pre-compression pressure) resulting in a final net LHV efficiency of 38.9 %.

Acknowledgment This publication has been produced with support from the BIGCCS Centre, performed under the Norwegian research program *Centres for Environment-friendly Energy Research (FME)*. The authors acknowledge the following partners for their contributions: ConocoPhillips, Gassco, Shell, Statoil, TOTAL, GDF SUEZ and the Research Council of Norway (193816/S60). The authors wish to thank Chao Fu (Norwegian University of Science and Technology) for providing the ASU energy consumption values.

392243 Overview of the U.S. Doe Carbon Capture Research and Development Program

Abstract:

The Department of Energy's Carbon Capture Program –managed within the Office of Fossil Energy and implemented by the National Energy Technology Laboratory– has been developing both second generation and transformational technologies that will enable carbon capture and storage (CCS) to become an effective and cost effective option for reducing carbon dioxide (CO2) emissions from fossil fuel based power plants. Successful research and development will enable CCS technologies to overcome the various technical, economic, and social challenges associated with this technology. The research being supported by the program leverages public and private partnerships to support the goal of broad, cost-effective carbon capture deployment. Fossil Energy is targeting demonstration of 2nd generation technologies that result in a captured cost of CO_2 less than $40/tonne in the 2020-2025 timeframe. Fossil Energy is also committed to extending R&D support to even more advanced transformational carbon capture technologies that will further increase competitiveness of fossil based energy systems beyond 2035. The program is supporting a portfolio of over 50 research projects form the laboratory to pilot scale which are designed to achieve these goals. These projects use a combination of developments in process chemistry, new chemical production methods, novel process equipment designs, new equipment manufacturing methods, and optimization of the process integration with other power plant systems (e.g., the steam cycle, cooling water system, carbon dioxide compression, etc.). This presentation will summarize the U.S. DOE's efforts to support the development of these carbon capture technologies and have them ready for commercial demonstration by 2020.

The Impact of RD&D on Market Opportunities for Power Plants with Carbon Capture

José R. Benítez, PE

Energy Analyst

Office of Program Performance and Benefits

Strategic Center for Coal

U.S. DEPARTMENT OF ENERGY | National Energy Technology Laboratory

Introduction

- **Who we are**
  - Office of Program Performance and Benefits (OPPB)
  - Located in the National Energy Technology Laboratory's (NETL) Strategic Center for Coal (SCC)
- **Our mission**
  - Conduct analysis which estimates expected benefits (the public return on investment) of successful Fossil Energy (FE)/NETL R&D programs
  - Support and inform policy and decision making
- **Our products**
  - Provide DOE FE/NETL management information to assist in communicating benefits of NETL R&D programs to key stakeholders (Congress, Office of Management and Budget (OMB), etc.) using suite of tools we have developed
    - FE/NETL CO_2 Saline Storage Cost Model
    - NETL/WVU Econometric Input-Output (ECIO) Model
    - FE NETL CO Capture Transport Utilization and Storage (CTUS)-NEMS
    - See http://www.netl.doe.gov/research/energy-analysis/analytical-tools-and-data

Overview

- Background
- Model Overview
- Comparison of NEMS-2014 Runs
- Example Analysis

Background

What is NEMS?

- NEMS is an acronym for the National Energy Modeling System, the custom made software developed by the Energy Information Administration (EIA) for nationwide energy market forecasting
- The results from NEMS are published yearly by EIA in its Annual Energy Outlook (AEO) reports
- NEMS is also used to study Congressionally directed scenarios, like the implementation of a Clean Energy Standard in service reports for Congress
- NEMS results are also often used to perform regulatory impact analyses by federal and state agencies
- Although NEMS is publically available, it is not generally run by outside entities. Still, its outputs provide valuable information for other models

5

Background

Previous Efforts

- **Old Paradigm (Pre-2012 Versions of NEMS)**
  - Carbon capture plants only considered a fixed cost of $10/tonne of CO_2 for transport and storage costs
  - Existing coal plant retrofits costs were based on an endogenous scaling of power plant costs
- **Model development began in 2009**
- **First version completed in 2010, represented transport and storage for new NGCC w/CCUS, and new and existing coal w/ CCUS .**
- **New Paradigm (NEMS-2012 and NEMS-2013)**
  - Plants with carbon capture were given the ability to consider region specific transportation and saline storage costs via the use of a carbon transportation pipeline network
  - Plant specific carbon capture retrofit costs were introduced

Carbon Transport and Storage (CTS) Model (Used in NEMS 2012 and 2013)

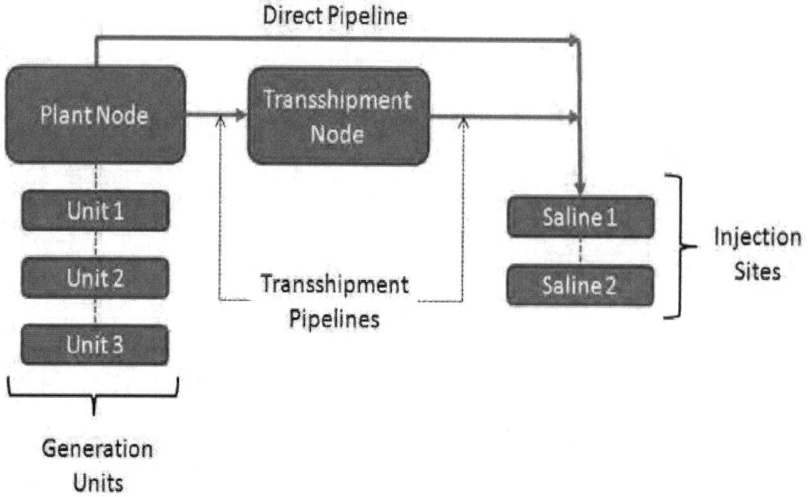

NETL

Background

Previous Efforts

- **Model expanded in 2012-2013 to more fully represent CO_2 EOR economics**
- **New Paradigm (NEMS – 2014)**
  - Electricity Market now "sees" the opportunity for utilizing the CO_2 being produced at CCS equipped power plants for EOR operations.
  - EOR operations now "see" a location and availability-dependent price of CO_2 from power plants.
  - LFMM modified to "see" the CO_2 opportunity in EOR and provide a location and availability dependent price of CO_2 from CTL facilities.

NETL

Carbon Transport, Utilization, and Transport (CTUS) Model (Used in NEMS – 2014)

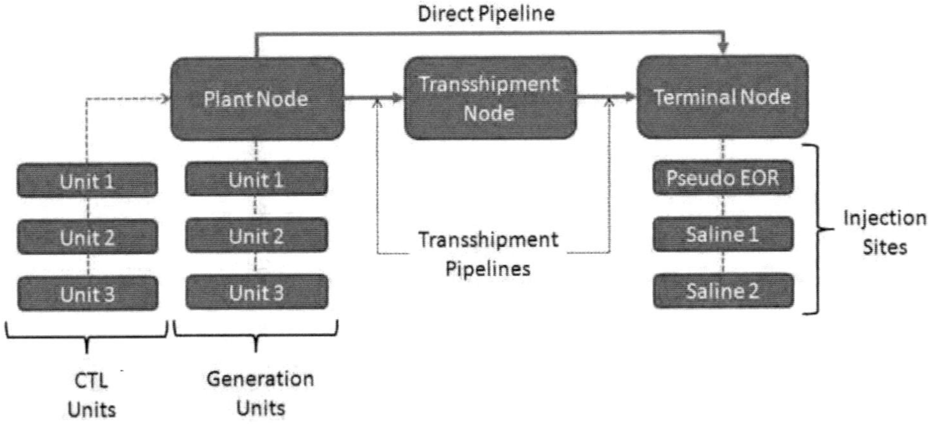

Model Overview

Model Technical Description

- **The FE/NETL CO_2 CTUS-NEMS Model**
  - The CTUS model is a NEMS plug-in sub-module written in GAMS (General Algebraic Modeling System)
  - Multiple sources, sinks, and EOR sites spatially represented
  - Estimates an optimal CCUS pipeline network (dedicated and trans-shipment) through Mixed Integer Programming (MIP) based on scenario evaluated
  - Passes transport and storage cost back to potential CCUS technology options in the main NEMS model which enables full energy economy representation in CCUS infrastructure simulations

Model Overview

NEMS General Structure

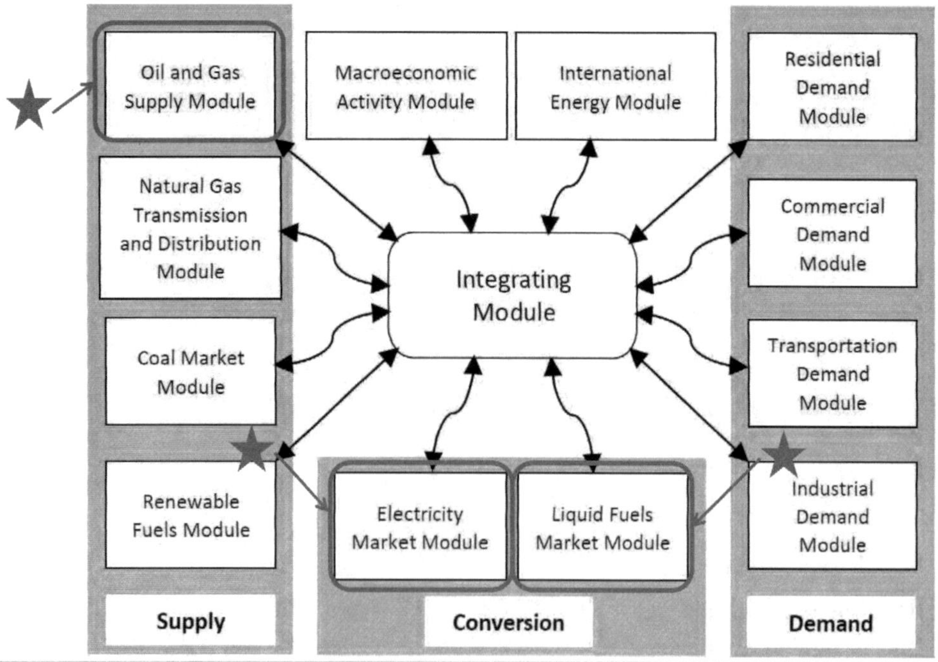

Model Overview

CO_2 CTUS NEMS Model Structure

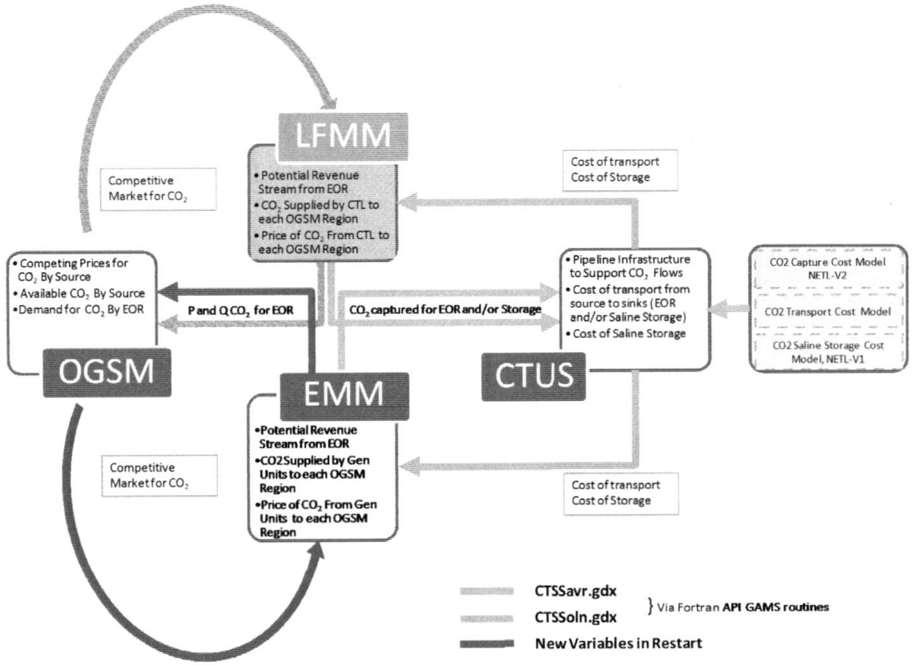

NETL

CTUS in NEMS-2014

NETL

Scenarios Evaluated

- **All scenarios use AEO 2014 final release as their basis**
  - EIA assumptions for research and development assumed
    - This includes 840 MW of carbon capture demonstration projects
    - It also assumes CCUS technology cost reductions over time due to domestic and international RD&D

Scenarios:

- **Reference Case (Reference)**
  - EIA AEO 2014 reference case
  - No modifications made
- **$10 Carbon Price Case ($10 Tax)**
  - EIA Greenhouse Gas $10 Carbon Tax Side Case (GHG10)
  - $10 Carbon tax introduced in 2015
  - Carbon tax escalates at 5% (real) per year
- **$15 Carbon Price Case ($15 Tax)**
  - $15 carbon price begins in 2020, escalated at 5% per year
  - No new nuclear builds until 2025
  - EIA low resource (low natural gas and oil) side case incorporated

NEMS Results: Capacity Additions

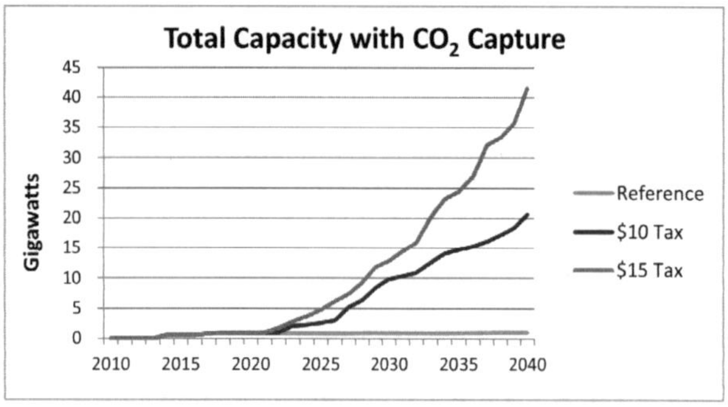

- **In the $10 tax case, slightly over 20 GW of CCUS deploys**
- **In the $15 tax case, over 40 GW of CCUS capacity deploys**

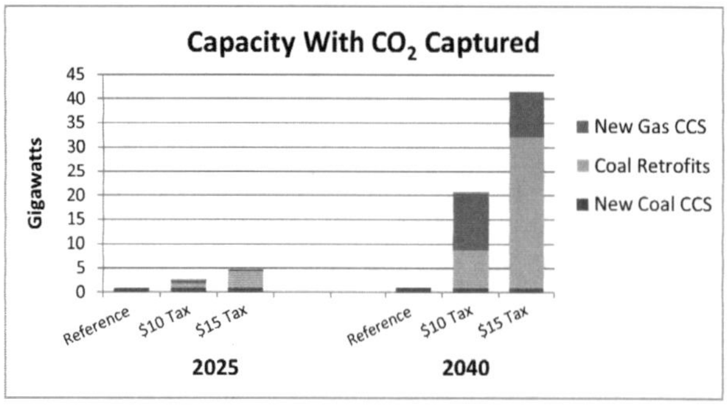

- **Capture from natural gas power plants dominates the $10 tax case**
- **In the $15 tax case, where natural gas is constrained (via the low resource case), coal retrofits dominate**

NEMS Results: Carbon Captured

- **Nearly 225 million metric tons of CO_2 is captured per year by power plants in the \$15 tax case**
- **Captured CO_2 comprises approximately 18% of the total reductions in CO_2 emissions (relative to the reference case CO_2 emissions) from the power sector in 2040**

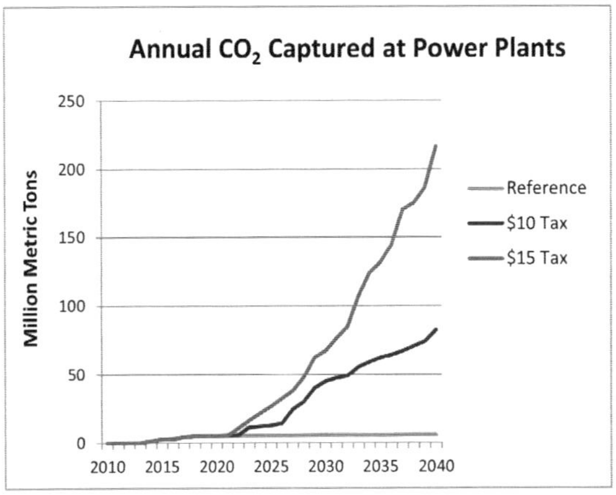

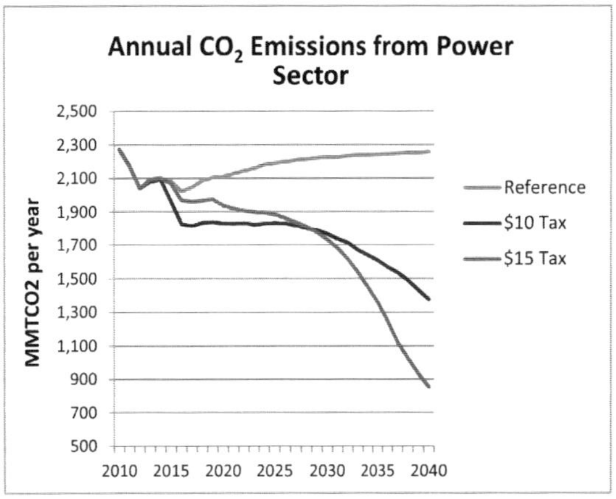

NEMS Results: Enhanced Oil Recovery

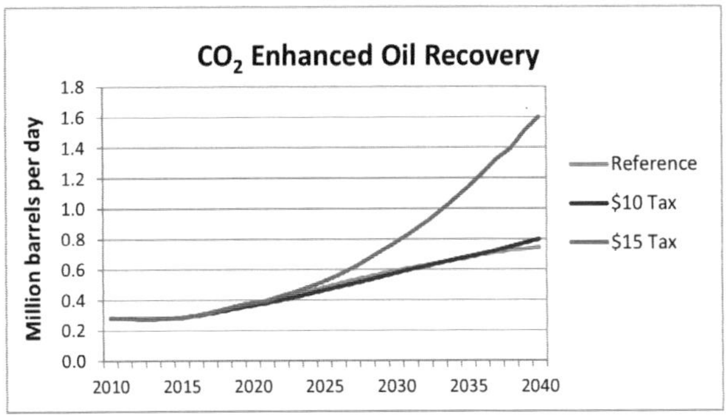

- **EOR enabled oil production doubles by 2040 in the \$15 GHG tax case**

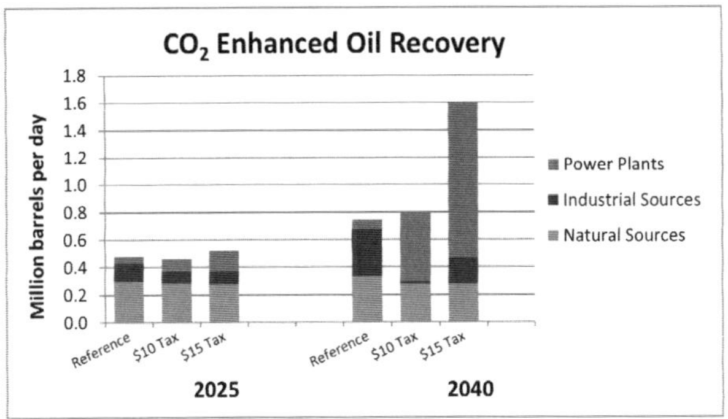

- **And the overwhelming majority of CO_2 for EOR is sourced from power plants**

Example Analysis

Example Analysis

Test Cases Evaluated

- **We built hypothetical Baseline and Capture Cost Reduction cases to test model functionality for use in benefits analysis**
- **All cases built off of EIA's AEO 2012 Reference Case – 4 cases ran and compared;**
  - Baseline Case
  - Baseline Case with Capture Cost Reduction
  - Baseline Case with Carbon Tax
  - Baseline Case with Carbon Tax and Capture Cost Reduction
- **For Baseline Case**
  - Removed part of EIA learning and other related assumptions
  - Removed demonstration plants
- **For Carbon Tax Cases**
  - Started tax at $15/tonne in 2013 and escalated by 5% per year :

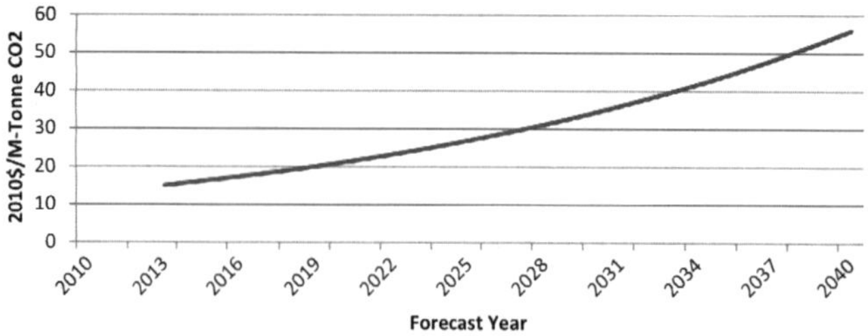

Example Analysis

Hypothetical Capture Cost Reductions

- **For Capture Cost Reductions**
  - The hypothetical Capture Cost Reduction cases incorporate a two-phase cost reduction across the model timeline.
  - Phase-1 cost reductions were introduced over a five year time period beginning in 2025, following a linear trajectory until the full phase-1 cost reduction is reached in 2030.
  - Between 2030 and 2034 improvements remain at the phase-1 reduction level.
  - Phase-2 cost reductions were introduced over a five year time period beginning in 2035, following a linear trajectory until the full phase-2 cost reduction is reached in 2040.

New Coal CCS Overnight Capital Cost

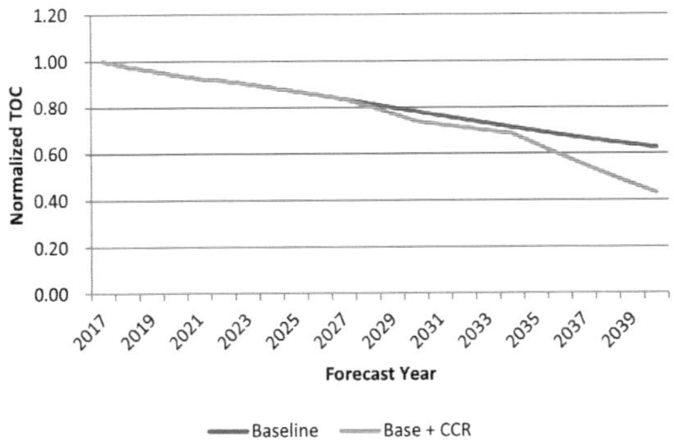

New Coal CCS Heatrate

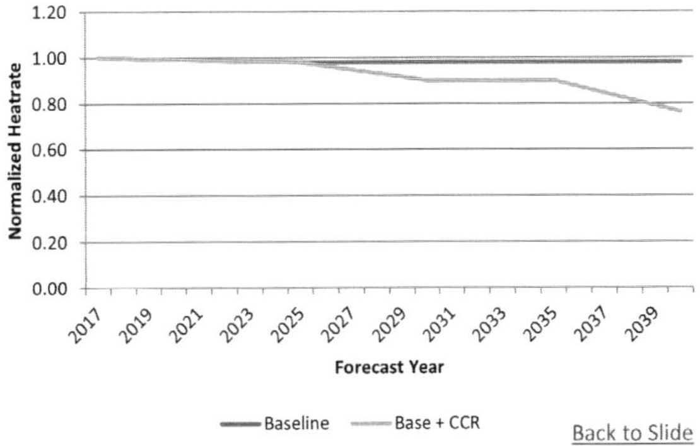

Back to Slide

NETL

Example Analysis

Capture Deployments Across Scenarios

Total Capacity with CO2 Capture

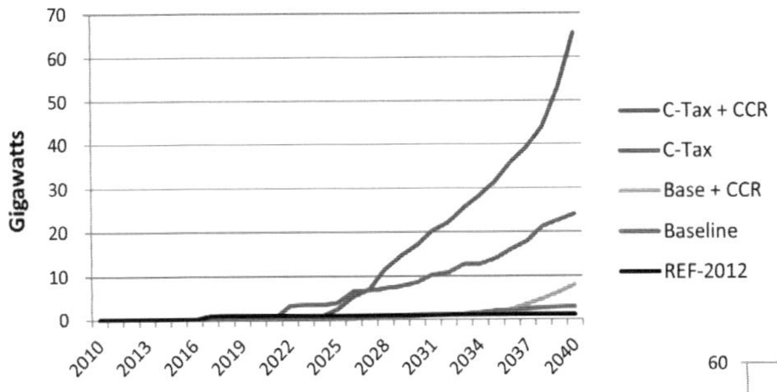

The second phase cost reductions result in a measurable increase in carbon capture in the last five years of the model time horizon

Taxing CO_2 emissions incentivizes substantially more power plant capture along with capture cost reductions, especially retrofits of coal capacity

Capacity With CO2 Captured

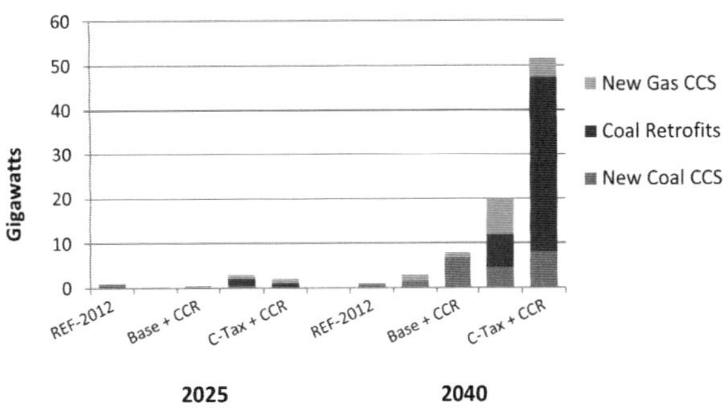

NETL

Example Analysis

CO2 EOR Oil Production

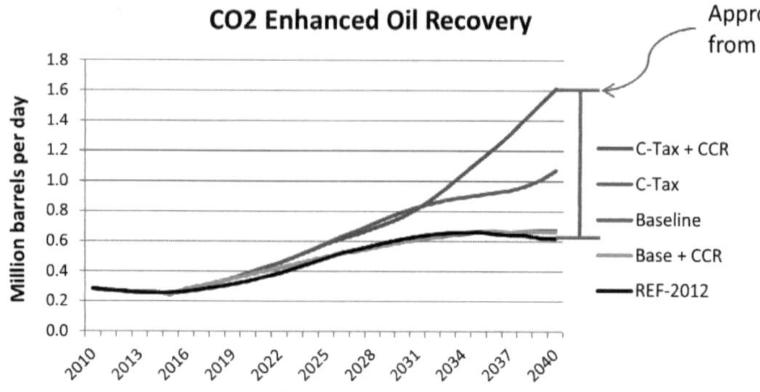

EOR production using CO_2 increases when more lower cost CO_2 becomes available

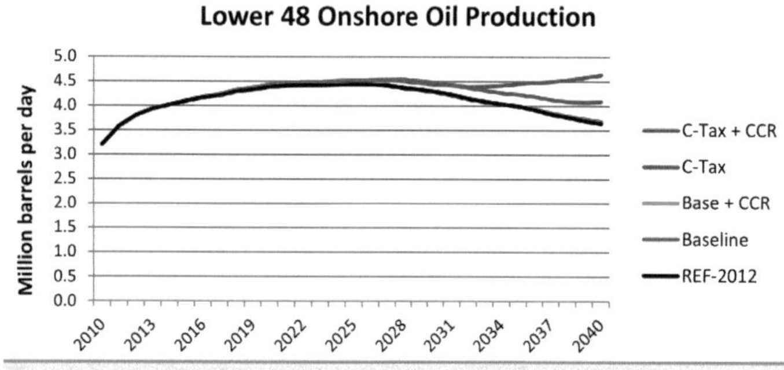

Overall oil production is higher as well leading to lower oil imports (along with lower demand resulting from the CO_2 fee)

Contact Information

- **Jose R. Benitez**
- jose.benitez@netl.doe.gov
- **(412) 386 – 7394**

377864 A Framework for Equation Based Optimization of Coal Oxycombustion Power Plants

Abstract:

In an oxy-fired power plant, coal is combusted in a N2 lean atmosphere, producing primarily CO2 and water. With a few additional processing steps, CO2 can be purified and compressed to supercritical conditions for utilization and/or sequestration. This approach may be more economical than post- and pre-combustion capture options for pulverized coal and IGCC power plants, respectively. When comparing these technologies optimized designs should be considered, lest a suboptimal design improperly disqualify a carbon capture approach.

From a systems perspective, optimizing an oxy-fired coal plant with carbon capture is very challenging, as the subsystems of the power plant are tightly integrated. For example, adjusting the flue gas recycle rate will impact the air separation unit (ASU), boiler, steam cycle and CO2 process unit (CPU) – essentially the entire power plant. Thus, it is inappropriate to optimize subsystems of the power plant individually. Instead, the entire process must be optimized at once to realize the tightest integration and best performance.

In this paper, we propose an equation-based framework for the optimization of coal oxycombustion power plants. By describing the entire process as equations instead of using commercial simulation software, we are able to use state-of-the-art nonlinear programming (optimization) algorithms capable of considering 100,000+ variables and constraints. The framework features several notable features:

- Embedded cubic equation of state thermodynamics with disappearing phases
- Simultaneous heat integration and process optimization
- Detailed distillation models with variable number of stages (important for ASU)
- Systematic initialization procedure
- Modular design for extension to other power systems and flowsheets

Using the framework, we designed cryogenic systems (ASU & CPU) for a coal oxycombustion power plant and conducted sensitivity analysis. For example, a Pareto optimality frontier was constructed by minimize the ASU specific separation energy for several O2 product purity specifications. These results provide other engineers vital information when selecting ASU operating conditions in oxy-fired power plants designs.

Progress in extending the framework to include a first-principles boiler model will also be discussed. We are developing a hybrid boiler model that provides 1-D resolution along the height of the radiant furnace for main flow and reaction related calculations, while radiative heat transfer calculations are resolved in 3-D. In contrast to CFD simulations, this model requires approximately 1 minute to evaluate, not days. Comparisons of this model with traditional CFD will be presented. We are also developing a trust-region based optimization framework to embed the hybrid model results via kriging interpolation into the full, equation-based power plant model. This is important, as it enables the use of efficient, large-scale nonlinear programming (optimization) algorithms.

References:

Dowling, A. W., & Biegler, L. T. (2013). Optimization-based Process Synthesis for Sustainable Power Generation. Chemical Engineering Transactions, 35, 1–12.

Dowling, A. W., & Biegler, L. T. A Framework for Efficient Large Scale Equation-Oriented Flowsheet Optimization. Submitted to Computers & Chemical Engineering.

378726 A Multivariable Model-Based Control Framework for a Post-Combustion CO2 Capture Plant

Abstract:

The future of fossil-fired power plants involves the reduction of Carbon dioxide (CO_2) emissions through the application of CO_2 capture and sequestration (CCS) processes. One of the most feasible CCS strategies to be integrated with different power plants, i.e., Integrated Gasification Combined Cycle (IGCC) and coal-fired combustion power plants, is the post-combustion CO_2 capture process using Mono-Ethanol-Amine (MEA) absorption. Therefore, the integration of CO_2 capture process and power plants will increase the need to design advanced model-based control systems to overcome the process interactions and maintain dynamic operability of power plant and CCS unit. Almost all the control schemes that have been recently proposed for the CO_2 capture process are focused on decentralized multi-loop control schemes [1-4]. A comprehensive model-based controller for the complete CO_2 capture process has not been addressed in the open literature. To the authors' knowledge, only one study have evaluated the controllability of different operational regions in a post-combustion CO_2 capture plant using 2x2 model-based multivariable controller [5].

In this study, an advanced multi-variable Model Predictive Control (MPC) scheme has been designed to evaluate the controllability of a post-combustion CO_2 capture unit in the presence of disturbances that may appear in the actual operation due to the plant's interactions with coal-fired power plants. The plant's dynamic model and the MPC-based control system were implemented in Aspen HYSYS and MATLAB, respectively. For the MPC-based control scheme, an instant client-server data communication link between the process simulator and the control framework was designed by the means of component objective model programing. The performance of the proposed MPC strategy for the CO_2 capture plant was evaluated under various scenarios and compared against the performance obtained by a decentralized multi-loop control scheme. The controllability study have shown that the proposed MPC-based control scheme rejected the disturbances faster with less deviation in the set points and did not violate the process constraints compared to the decentralized control scheme. In order to provide insight into the operation of a CO_2 capture plant, the proposed MPC-based control strategy and the CO_2 capture plant were embedded within a dynamic multi-objective optimization framework to search for the optimal scheduling of CO_2 capture in the presence of sinusoidal disturbances in the flue-gas flow rate. Two different scenarios were considered in the optimal scheduling problem: i) low CO_2 emission scenario where total CO_2 capture is reduced by 2.3%; ii) high electricity generation scenario on which the reboiler unit was only allowed to extract 37% less energy than that used for the base case scenario. The insight gained from the optimal scheduling study will support the design of optimal operating policies for the integration of CO_2 capture to coal-fired power plants by considering the dynamic performance of the system in closed-loop.

References

[1] Thanita Nittaya, Peter L. Douglas, Eric Croiset, Luis A. Ricardez-Sandoval, "Dynamic modelling and control of MEA absorption processes for CO_2 capture from power plants," *Fuel* , 2014;116:672–691.

[2] Panahi M, Skogestad S. , "Economically efficient operation of CO_2 capturing process part I: self-optimizing procedure for selecting the best controlled variables," *Chem Eng Process* , 2011;50(3):247–53.

[3] Lawal A, Wang M, Stephenson P, Koumpouras G, Yeung H. , "Dynamic modelling and analysis of post-combustion CO_2 chemical absorption process for coal- fired power plants," *Fuel,* 2010;89(10):2791–801.

[4] Lin Y, Pan T,Wong DS, Jang S, Chi Y, Yeh C., "Plantwide control of CO_2 capture by absorption and stripping using monoethanolamine solution," *Ind Eng Chem Res,* 2011;50(3):1338–45.

[5] Panahi M, Skogestad S., "Economically efficient operation of CO_2 capturing process part II: control layer," *Chem Eng Process,* 2012;52:112–24.

380816 Thermodynamic and Economic Analysis of an Integrated Gasification Combined Cycle (IGCC) Plant with Various Carbon Capture Technologies

Abstract:

The levels of CO_2 emitted by coal combustion raise enormous concerns about its utilization as energy source because of the copious effects on global climate. In this scenario, carbon capture technologies (CCT) emerge as the effectual remediation processes to curtail CO_2 emissions from Coal power plants. An IGCC (integrated gasification combined cycle) is a widely used electrical power generation system that allows for a variety of feedstocks with high efficiencies. And the integration of CCT into IGCC requires a demonstration to commercial scale to minimize technological risks, capital and operational cost.

In this study, a 300 MW class IGCC plant without CCT was simulated using the PRO/II software package, and thermodynamic analysis was performed as a first step. The simulated results were compared to the basic design data for a 300 MW Class IGCC demonstration plant to evaluate the validity. And the effects of coal grade on the power generation and consumption in viewpoint of overall efficiencies through Sankey diagrams for the energy and exergy flow associated with the first and second laws of thermodynamics with major components of the IGCC.

As a second step, this paper evaluated from the technical to economic points of view via process modeling and simulation of five carbon capture technologies involving: (i) a-MDEA, (ii) Retisol, (iii) Selexol, (iv) Purisol and (v) Cryogenic process. Though the crucial choice of the technology to actualize favors an a-MDEA CCT when emphasis is on cost minimization, the technical evaluation favors a Selexol CCT.

DYNAMIC PROCESS MODELING OF A SUSTAINABLE COAL, WIND, AND SMALL MODULAR REACTOR HYBRID ENERGY SYSTEM

Kyle L. Buchheit and Joseph D. Smith, Ph.D.

Missouri University of Science and Technology, Rolla Missouri

ABSTRACT

The availability of cheap electricity is one of the biggest factors for improving quality of life. With the debate on the effects of carbon dioxide emissions continuing, several countries have either implemented or are considering the reduction of emissions through various economic means. The inclusion of a monetary penalty on carbon emissions would increase the prices of electricity produced by carbon-based sources. The push for large-scale renewable sources of energy has met problems with regards to energy storage and availability. The proposed coal, wind and nuclear hybrid energy system would combine a renewable energy source, wind, with traditional and stable energy sources, coal and nuclear, to create an integrated and sustainable system. Instead of a light water reactor, a next generation small modular nuclear reactor will be evaluated. The coal system will use a pressurized circulating fluidized bed system, which can not only take coal, but also biomass as a carbon feedstock. This system also employs a high temperature steam co-electrolysis unit for the utilization of carbon dioxide emissions for the production of synthetic gas which can be used in the production of transportation fuels or chemicals.

A rigorous dynamic process model of the system will be constructed. This model will allow for proper scaling of each individual component in the overall hybrid system along with the dynamic characteristics inherent in a fluctuating wind system. Increased integration of the water and heat cycles will be evaluated for improved efficiency. A detailed Aspen Plus user model will be utilized to demonstrate the feasibility of the main hybrid energy system components. Aspen Dynamics and MatLAB / Simulink modeling will integrate the main process model with the dynamic wind model to look at overall sustainability, power generation, and chemical production.

387253 Cost of Electricity of a Staged, Pressurized Oxy-Combustion Power Plant

Abstract:

The immediate need for low cost carbon capture processes has prompted the recent development of pressurized oxy-combustion processes [1-3]. At increased combustion pressure, dew point of flue gas moisture is increased, allowing for effective integration of the latent heat of flue gas moisture condensation into the Rankine cycle. This increases the net plant efficiency and reduces costs. The Staged, Pressurized Oxy-Combustion (SPOC) process further enhances the efficiency and reduces costs by reducing the recycle of flue gas to near zero. This is achieved by staging the delivery of fuel and, thereby, using excess oxygen along with the combustion products of previous stages as diluents to manage the temperature and heat flux during the combustion process. Compared to first generation oxy-combustion, SPOC process results in an increase in net plant efficiency of about 5 percentage points.

The cost of a 550 MW_e SPOC power plant, modeled in ASPEN Plus, with 90% carbon capture is described. The effect of other factors such as contingencies, cost of CO_2 and financial structure, etc. on the cost of electricity is analyzed. Results show that in comparison to an air-fired combustion power plant without carbon capture, the SPOC process can limit the increase in cost of electricity to less than 35%, in line with the US Department of Energy's goal.

Acknowledgements

- This material is based upon work supported by the Department of Energy under Award Number DE-FE0009702.
- Ameren advising team: Rich Smith, Tom Callahan, George Mues.
- Engineering, consulting, and cost estimation services provided by Burns & McDonnell.

References:

1. Gopan, A., et al., *Process design and performance analysis of a Staged, Pressurized Oxy-Combustion (SPOC) power plant for carbon capture.* Applied Energy, 2014. **125**(0): p. 179-188.
2. Hong, J., *Techno-economic analysis of oressurized oxy-fuel combustion power cycle for CO2 capture.* 2009, MIT: Cambridge.
3. Zheng, L., et al. *Optimization of a high pressure oxy-fuel combustion process for power generation and CO2 capture* in *International Technical Conference on Coal Utilization & Fuel Systems.* 2010. Clearwater, Florida, USA.

Brief Review of Catalytic Oxidative Coupling of Methane to Ethane and Ethene

Evan J. Granite

National Energy Technology Laboratory

Pittsburgh, PA 15236-0940

Abstract

Methane, the simplest hydrocarbon, is the major constituent of natural gas. Much of the methane produced worldwide is flared due to lack of pipelines. The prime use of methane is as a fuel for home heating and cooking, as well as electricity production. Because of its abundance, methane has the potential of becoming a major feedstock for petrochemical production. The problem is that methane is a stable molecule, and the direct conversion to useful chemicals is difficult.

In 1982 Keller and Bhasin[1] reported that many metal oxides can catalyze the oxidative coupling of methane to the more valuable chemicals ethane and ethene. In the presence of the catalyst, methane reacts with oxygen and forms the higher hydrocarbons ethane and ethene during oxidative coupling, shown in reactions (1) and (2) below:

$$2\ CH_{4\,(gas)} + \tfrac{1}{2}\ O_{2\,(gas)} \rightarrow C_2H_{6\,(gas)} + H_2O_{(gas)} \quad (1)$$

$$2\ CH_{4\,(gas)} + O_{2\,(gas)} \rightarrow C_2H_{4\,(gas)} + 2\ H_2O_{(gas)} \quad (2)$$

The catalytic deep oxidation of methane to carbon monoxide and carbon dioxide will also occur, shown as reactions (3) and (4) below:

$$CH_{4\,(gas)} + 3/2\ O_{2\,(gas)} \rightarrow CO_{(gas)} + 2\ H_2O_{(gas)} \quad (3)$$

$$CH_{4\,(gas)} + 2\ O_{2\,(gas)} \rightarrow CO_{2\,(gas)} + 2\ H_2O_{(gas)} \quad (4)$$

Unfortunately the yields of ethane and ethane obtained by Keller were small (5% maximum total C_2 yield) and the process temperature required was very high (800°C). In the ensuing thirty two years, a vigorous search proceeded for an active coupling catalyst[2-23].

The first breakthrough occurred in 1985, when Lunsford[2] reported 19% C_2 yields when Li doped MgO was used as a catalyst. In 1986 Otsuka[3,4] reported rather spectacular C_2 yields of 30% with LiCl-MnO_2 and LiCl - NiO catalysts. In this case however, the catalyst was short lived, suggesting that chlorine in the catalyst is a reagent, and chlorine is known to inhibit the deep oxidation of methane.

Labinger[18] suggested in 1988 that there is a fundamental limit on the C_2 yield of 30%. The catalysts typically require high operating temperatures of 600 - 800°C. Unfortunately these factors work against commercialization of the catalysts.

In order for an oxidative coupling catalyst to be commercially viable, three criteria must be met[18]. First a total C_2 yield (preferably as the more valuable ethene) of at least 30% is needed. Second the process should operate at lower temperatures to be less energy intensive. Finally the catalyst should have a long life. None of the metal oxide catalysts examined exhibits all of the necessary characteristics to be commercially viable. The discovery of such an oxidative coupling catalyst remains an elusive and sought after prize. The mechanisms by which the catalysts convert methane will be discussed and future research directions will be suggested.

Disclaimer

Literature Cited

1. Keller, G.E. and Bhasin, M.M., J. of Catalysis, 73, pp. 9-19, 1982.
2. Lunsford, J. and Ito, T., Nature, 314, p.721, 1985.
3. Otsuka, K.; Jinno, K.; and Morikawa, A., J. of Catalysis, 100, pp. 353-359, 1986.
4. Otsuka, K.; Liu, Q. and Morikawa, A., J. Chem. Soc., Chem. Commun., pp. 586-587, 1986.
5. Lunsford, J.; Wang, J-X.; and Lin, C-H., J. of Catalysis, 111, pp. 302-316, 1988.
6. Baerns, M. and Bytyn, W., German patent DE 3,406,751, applied 24 Feb. 1984, granted 29 Aug. 1985, 32pp.
7. Imai, H.; Tagawa, T. and Kamide, N., J. of Catalysis, 106, pp. 394-400, 1987.
8. Deboy, J. and Hicks, R., J. of Catalysis, 113, pp. 517-524, 1988.
9. Iwamatsu, E.; Moriyama, T.; Takasaki, N. and Aika, K. J. of Catalysis, 113, pp. 25-35, 1988.
10. Gaffney, A.; Jones, A.; Leonard, J. and Sofranko, J.A. J. of Catalysis, 114, pp. 422-432, 1988.
11. Lane, G.; Miro, E.; and Wolf, E., J. of Catalysis, 119, pp. 161-178, 1989.
12. Agarwal, S.K.; Migone, R.A.; and Marcelin, G., J. of Catalysis, 121, pp. 110-121, 1990.
13. Miro, E.; Santamaria, J.M.; and Wolf, E.E., J. of Catalysis, 124, pp. 451-476, 1990.

14. Aika, K.; Fujimoto, N.; Kobayashi, M.; and Iwamatsu, E., J. of Catalysis, 127, pp. 1-8, 1991.
15. Becker, S. and Baerns, M., J. of Catalysis, 128, pp. 512-519, 1991.
16. Choudhary, V.R. and Rane, V.H., J. of Catalysis, 130, pp. 411-422, 1991.
17. Papageorgiou, D.; Vamvouka, D.; Boudouvas, D.; and verykios, X.E., Catalysis Today, 13, pp. 391-400, 1992.
18. Labinger, J. Abstracts of Papers, 201st ACS National Meeting, p. 7, April 1991.
19. Baronetti, G.; Padro', C.; Scelza, O.; and Castro, A., Applied Catalysis A: General, vol. 101, pp. 167-183, 1993.
20. Ruckenstein, E. and Khan, A., Catalysis Letters, vol. 18, pp. 27-35, 1993.
21. Zhou, S.; Zhou, X.; Wan, H.; and Tsai, K., Catalysis Letters, vol. 20, pp.179-183, 1993.
22. Sinev, M. Yu; Fattakhova, Z.T.; Lomonosov, V.I.; Gordienko, Yu. A., Journal of Natural Gas Chemistry, vol. 18, 273-287, 2009.
23. Alexiadis, V.I.; Thybaut, J.W.; Kechagiopoulos, P.N.; Chaar, M.; and Van Veen, A.C., Applied Catalysis B: Environmental, vol. 150-151, pp. 496-505, 2014.

Reaction Studies on Na_2WO_4-Mn/SiO_2 Catalyst for Oxidative Coupling of Methane

Aseem[a], Geofrey Goldwin[b], Marlon Conato[a], Jeffrey D. Rimer[a], Michael P. Harold[a] and Ken Blackmon[b]

[a] *University of Houston, Department of Chemical and Biomolecular Engineering, Houston, TX 77204*
[b] *Total Petrochemicals & Refining USA, Inc., LaPorte, TX 77571*

The oxidative coupling of methane (OCM) to higher molecular weight hydrocarbons has great appeal as an alternative route to chemicals and polymers. The reaction system is challenging because of large differences in the reactivity of the reactant methane and desired products (e.g., ethylene, ethane, etc.) at the temperatures needed to activate methane. Moreover, the reaction system is highly exothermic and is noted for significant parametric sensitivity.

A large number of systems have been studied over the past few decades, but a catalyst and reactor concept has not been found that give sufficiently high C_{2+} yield. The most studied catalysts to date for the OCM reaction system have been alkali-promoted alkaline earth metal oxides, transition metal oxides and rare earth metal oxides. Na_2WO_4-Mn/SiO_2 catalyst has attracted a significant amount of attention because of its favorable catalytic performance. In this study we evaluate the activity of the catalyst in particulate and washcoat form using both fixed-bed and monolith reactor configurations. We also carry out a parametric study to quantify the effects of residence time, reaction temperature, oxygen concentration, and dilution.

The catalyst was prepared by the incipient wetness impregnation method and its structural properties were characterized by a variety of analytical techniques (e.g., XRD, BET, ICP and SEM). The prepared catalyst has a surface area of 2.6 m^2/g and was sieved to a size of 50-70 mesh. Residence time variations were accomplished by changing the weight of catalyst while keeping the flow rate constant, thereby avoiding changes in flow patterns inside the reactor. Figure 1 shows the dependencies of methane conversion, C_2 selectivity, and C_2 yield on residence time at 800°C. The conversion of methane increased with increase in residence time but the C_2 selectivity decreased due to deep oxidation in the reactor with prolonged exposure to the catalyst.

We observed that, with higher CH_4/O_2 ratios and concentrated feed, the reactor system becomes quite sensitive to the operating conditions. In particular, the steep axial temperature limits the maximum oxygen concentration that can be used. By diluting the feed, the reaction can be conducted at higher CH_4/O_2 ratios. We will also show the effect of dilution of catalyst with inert and catalyst structure (i.e. particles vs. washcoat) on the product compositions. Washcoating is of particular interest since it reduces the contact time of C_2 hydrocarbons within the internal porous structure of the catalyst, thereby inhibiting deep oxidation.

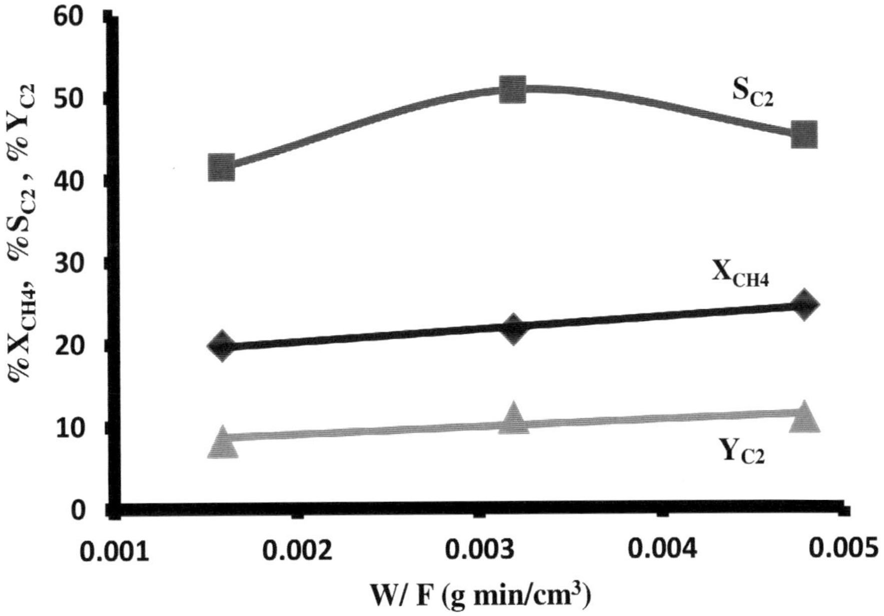

Figure 1. Dependence of CH_4 conversion, C_2 selectivity, and C_2 yield on space time at T = 800^0C and a CH_4:O_2 ratio of 3:1.

2014 AIChE Annual Meeting
November 16th - November 21st, 2014
Hilton Atlanta
Atlanta, GA

Abstract Submission

PROGRAM TOPIC:
TE002 Natural Gas Reaction Engineering

EXACT TITLE OF PAPER:
Stabilization of ZnO in ZSM-5 for Methane Dehydroaromatization

Presenting Author: Yungchieh Lai
Department of Chemical engineering, University of Pittsburgh, Pittsburgh, PA
Contact Information: yul69@pitt.edu

Co-Authors: Götz Veser
Department of Chemical Engineering, University of Pittsburgh, Pittsburgh, PA
Contact Information: gveser@pitt.edu

Abstract:

The low cost of natural gas and the vast amount of accessible shale gas reserves in the US has made production of the most major aromatic product, benzene, via catalytic dehydroaromatization (DHA) of methane an economically attractive pathway. A direct conversion of methane into benzene is given by 6 $CH_4 \rightarrow 9\ H_2 + C_6H_6$ and proceeds over zeolite-based catalysts at ~700-800^{0}C with good selectivities (>60%) but low conversion. The most popular catalysts for this reaction are bi-functional Mo/zeolite systems in which the metal site activates methane, followed by oligomerization of the methyl to benzene on the Bronsted Acid Sites (BAS) of the zeolite. The good selectivity for the desired product benzene is attributed to the zeolite pore size which is similar to the benzene diameter (~ 6 angstrom). However, similar to other microporous zeolite materials, poor transport of primary reaction products result in rapid deactivation of the catalyst via coking. While this coke formation is reversible via oxidative regeneration of the catalyst, it results in an inactive oxidized form of the metal which then has to undergo a lengthy activation period to (re)form the active (oxy-carbide) phase. These issues motivate the search for more easily regenerable catalysts for this reaction.

ZnO/HZSM-5 has been studied intensively for DHA of light alkane (C_2 ~C_4) and was shown to be very active. Zn is active in these reactions as it is associated to BAS with oxidation state order of two, which can be in the form of isolated Zn^{+2}, binuclear $[ZnOZn]^{+2}$ clusters, or larger clusters of zinc oxide [1]. However, few studies used ZnO/HZSM-5 for methane DHA to-date, although ZnO/HZSM-5 is a promising candidate which may not only show activity for this reaction but may also not require the lengthy activation period after oxidative regeneration of a deactivated catalyst that plagues current Mo-catalysts. However, metallic Zn has a very high vapor pressure so that ZnO is quickly lost at high-temperature conditions, with only Zn^{+2} stabilized in zeolite framework surviving [2]. Hence, only a limited amount of Zn can be retained after reduction at high temperature (Zn/Al$\leq$ 0.5), and enhancement of ZnO stability for this catalyst is necessary to use it for methane DHA.

Here, we report a novel pathway to prepare ZnO@ZSM5 nanocomposite catalysts in which ZnO nanoparticles are embedded in ZSM-5. The encapsulation can result in a much better retention of ZnO in a high temperature reducing environment compared to conventional ZnO/ZSM5 made by impregnation method. Specifically, we find that while some Zn was lost rapidly during high-temperature exposure to a reducing atmosphere, the remaining Zn was stable even after long-time exposure (5%H_2 for 96 hr at 700^0C). We will present a detailed characterization of this nanocomposite, and discuss the stable population of Zn in this structure as well as reactivity data in the performance of this catalyst for methane conversion.

Reference:

1. Almutairi, S.M., et al., *Structure and reactivity of Zn-modified ZSM-5 zeolites: the importance of clustered cationic Zn complexes.* ACS Catalysis, 2011. **2**(1): p. 71-83.
2. Biscardi, J.A., G.D. Meitzner, and E. Iglesia, *Structure and density of active Zn species in Zn/H-ZSM5 propane aromatization catalysts.* Journal of Catalysis, 1998. **179**(1): p. 192-202.

2014 AIChE Annual Meeting

November 16th - November 21th, 2014

Hilton Atlanta

Atlanta, GA

Abstract Submission

Program Topic: TE002 Natural Gas Reaction Engineering

Title: Oxidation of methane and ethane to oxygenates under mild aqueous conditions using ZSM-5 catalysts

Presenting Author: Graham J Hutchings[§]

Presenting Author Email: Hutch@cardiff.ac.uk

Co-authors:

Ceri Hammond[§], email: HammondC4@cardiff.ac.uk

Michael Forde[¥], email: michael.forde@sta.uwi.edu

Nikoloas Dmitratos[§], email: DimitratosN@cardiff.ac.uk

[§] Cardiff Catalysis Institute, School of Chemistry, Cardiff University, Main Building Park Place, Cardiff, CF103AT , United Kingdom. Tel.: +442920874059

[§], [¥] Department of Chemistry, University of the West Indies, St. Augustine Campus, Trinidad and Tobago. Tel: +18686622002 ext83544

Abstract

The valorisation of natural gas components, *i.e.* conversion of methane and ethane to more reactive oxygenated products, under mild environmentally benign conditions has not been realised to date. This is due to difficulty in functionalising these highly inert hydrocarbons without further converting the primary products to deeper oxygenates such as carbon oxides, which are undesirable. Notably, several routes to producing alcohols from lower hydrocarbons have been described but none have found commercial viability and many of the chemistries involved are not in accord with the current global trend towards green chemical technologies.

We have found that ZSM-5 materials containing trace amounts of iron can catalyse the conversion of methane and ethane to oxygenated products at appreciable yield (10% and 50% conversion in batch process respectively) and high selectivity (<5% carbon oxides) under mild aqueous conditions using hydrogen peroxide as the oxidant. The catalyst can be tuned to selectively produce methanol from methane by the incorporation of copper into the catalyst system. Interestingly, this modification allows for the direct production of ethylene at appreciable levels of conversion from ethane at 50°C in aqueous phase. Though our characterisation and catalytic studies have shown that the catalytic iron sites in these materials are extra-framework iron species, we have elucidated different mechanistic pathways in the oxidation of methane and ethane over these materials. We will also present data that demonstrate the possibility of incorporation of molecular oxygen into the products which opens new avenues to utilisation of the preferred oxidant, oxygen. These significant data show the potential of these materials to be employed in natural gas valorisation processes.

「Development of a two beds circulating fluidized bed reactor system for non-oxidative aromatization of methane over Mo/HZSM-5 catalyst]

Zhanguo, Zhang, Yuebing Xu, Yang Song and Yoshizo Suzuki

National Institute of Advanced Industrial Science and Technology (AIST),
Onogawa 16-, Tsukuba 305-8569, Japan
z.zhang@aist.go.jp

1. Introduction

The non-oxidative aromatization of methane provides a promising route of direct methane conversion to valuable aromatics and has been extensively studied from both fundamental and applicative points of view [1,2]. Mo/HZSM-5 has been well recognized the most promising catalyst for the reaction, but it suffers from severe deactivation due to coking [3,4]. Adding a small amount of an oxygen-containing gas, such as CO2 or H2O, into CH4 is effective for slowdown of the coking rate [5, 6], but all coking reactions cannot be fully inhibited by this approach. Thus it is essential to develop a highly efficient, continuous catalyst regeneration technology for industrialization of the title catalytic reaction system. Over the past thirteen years we have been focusing our effort on the development of a really applicable technology for regeneration of coked Mo/HZSM-5 catalyst. Sustained effort has brought us the following findings of practical significance [7-10]. 1) The high initial activity of Mo/HZSM-5 can be maintained via its cyclic regeneration in H_2 at 973-1073 K, and the H_2-regeneration approach is now well recognized to be the only practical way of reactivating coked Mo/HZSM-5 catalyst. 2) Continuous H_2 regeneration can be easily realized in a two beds type of circulating fluidized bed reactor and the high initial activity of Mo/HZSM-5 can be maintained at 1073 K and a high space velocity up to 10000 mL/g/h for several tens hours without replenishment of fresh catalyst. 3) A binder-free, well shaped fluidizable Mo/HZSM-5 has been developed and successfully tested in a two beds type of circulating fluidized bed reactor system. In this presentation we will give a detailed description of how the major operating parameters are determined including temperature, space velocity and the mean residence times of the catalyst in both methane convertor and catalyst regenerator, and present some typical results we obtained under continuous catalyst regeneration operation mode.

2. Experimental

2.1 Catalyst

A binder-free, spherical-shaped 6%Mo/HZSM-5 catalyst having a compressive strength of about 850 kgf/cm^2 was used for this study. Fig. 1 shows a SEM image of this catalyst. It was supplied by Meidensha. Co, based on a ball-milled, nanosized HZSM-5 sample (the parent zeolite is a commercial product of Tosoh Co. and has a Si/Al of 20) and manufactured by spray-drying technique. It has an average particle size of about 160 μm and a minimum fluidization velocity of about 0.8 cm/s [7]. For comparison a 17 wt% binder-added, cylinder-shaped 6%Mo/HZSM-5 (supplied by Meidensha Co.) and a self-prepared 6%Mo/HZSM-5 powder catalyst were also tested in this study [8]. The powder catalyst was prepared using a conventional impregnation method and the zeolite was purchased from Zeolyst International (CBV3024) and has a Si/Al ratio of 15. The Mo-impregnated powder catalyst precursor was dried in air at 383 K over night and further calcinated at 773 K for 5 h. The obtained powder catalyst was pressed, crushed and sieved to particles in sizes of 250-500 μm for use.

2.2 Activity evaluation

Activity evaluation tests in continuous catalyst-regeneration mode were conducted in a two beds type of circulating fluidized bed reactor system at 1073 K, space velocities of 2000-3500 mL/g/h and catalyst-circulation rates of ~1.7g/min. The reactor system mainly consists of a quartz-made CH_4 convertor, a quartz regenerator, a glass riser, a glass cyclone, and a steel-made screw feeder. Fig. 2 shows the configuration. Two gold furnaces were used to heat CH_4 convertor and regenerator, which allowed visual observation on fluidization of catalyst particles in the two reactors at the reaction temperature of 1073 or 1123 K. Uniform bubble fluidized beds were confirmed to form in both CH4 convertor and regenerator at the test conditions. The reacted stream was sampled at the outlet of the CH4 convertor and analyzed using two GCs and therefore CH_4 conversion and benzene formation rate were monitored. The rate of catalyst circulation between the two reactors was controlled by the screw feeder, the rate of which was varied in each test according to the catalytic performance of catalyst. The bed height of the CH_4 convertor was controlled by the vertical position of the down-end of the riser inserted in the convertor. The up-end of the riser was connected to the cyclone, where most of catalyst particles contained in the reacted stream and lifted up from the convertor was recycled and then transported to the regenerator for their regeneration in H2. Sufficiently regenerated particles were then flowed down to the feeder and fed back to the CH_4 convertor.

To determine the operation parameters for the tests in the circulating reactor system, a micro-fixed bed reactor with an i.d. of 8 mm and a micro-fluidized bed reactor having an i.d. of 15 mm were also used in this study. Their detail configurations and the procedures for the tests in these two reactors were given in previous papers [7-9]. The operation parameters tested included temperature, space velocity and the mean residence times of catalyst in both CH_4 convertor and regenerator.

3. Results and discussions

Fig. 3 compares the catalytic performances of the two well-shaped catalysts and 100% HZSM-5-based powder catalyst, which were obtained in the micro fixed-bed reactor at 1073 K and 10,000 mL/g/h under the periodic 5 min CH_4–10 min H_2 switch operation mode. While addition of a 17wt% binder material to Mo/HZSM-5 led to an obviously decreased performance and made the catalyst less stable, the binder-free catalyst showed almost the same high activity and stability as the self-prepared, 100%HZSM-5-based powder catalyst. This strongly suggests the high potential of the finder-free catalyst for use in any practical fluidized bed reactor system.

Then the binder-free catalyst was tested in the same fixed bed reactor to examine the effect of reaction temperature on its activity stability. As shown in Fig. 4, the catalyst's CH_4 conversion activity increases whereas its stability decreases with increasing temperature. XRD and TG Measurements of the spent samples further confirmed that the more rapid deactivation xwas resulted from more severe coke formation, not from breakdown of the zeolite framework. This observation suggests that the optimal operation temperature of the test catalyst should be around 1073 K to obtain a possibly high stability.

In addition to operation temperature, space velocity is another important parameter to be determined for designing a practical reactor system. Some tests for optimization of reaction space velocity were first conducted in the micro fixed bed reactor at 1073K. The space velocity was varied from 3000 to 60000 mL/g/h and the results showed that an identical maximum benzene formation rate could be reached in a low range of < 10000 mL/g/h. Out of this range the maximum benzene concentration in the reacted stream decreased with increasing space velocity, suggesting a decreasing CH4 conversion. Note that a possibly high one pass CH4 conversion is required to maximize the performance of any real process. Any lower CH4 conversion at a higher space velocity (> 10,000 mL/g/h) could lower the process performance and therefore should be avoided. Then confirmation on the optimal space velocity was further made by performing tests in the micro fluidized bed reactor at the same temperature. As shown in Fig. 5, at the space velocity 11000 mL/g/h the maximum benzene concentration in the reacted stream reached only about 80 of the stable level reached at 3000 mL/g/h, while it attained about 91% at the space velocity 5000 mL/g/h. This suggests that in fluidized bed operation mode, the optimal space velocity for the test catalyst should be in a range of 3000-11000 mL/g/h.

Moreover, for designing and operating a two beds type of circulating fluidized bed reactor system, determination of a pair of proper mean residence times of catalyst particles in its CH4 convertor and regenerator is needed. Most of the tests for this purpose were conducted in the micro fixed bed reactor in periodic CH4-H2 switch operation mode [9]. The results obtained suggested that rather than shortening the exposure duration of Mo/HZSM-5 catalyst to CH4 (reaction time) extending its contact time with H2 is more effective for recovering its activity and maintaining its stability. Fig. 6 shows the results from a typical test in periodic regeneration mode (periodic CH4-H2 switch operation mode). The data were collected with the finder-free catalyst and the test was conducted in the micro fluidized bed reactor under a periodic CH4-H2 switch operation mode (CH4 10min- H2 40 min. The total CH4 feed time (reaction time) in the 5 days' test was about 1900 min. The catalyst slightly deactivated in each day's periodic operation because of a slow accumulation of coke, but its activity was well recovered after an additional 1 h regeneration at 1073 K in the next day's test. This proves that the test binder-free catalyst enables a stable benzene production at 1073 K as long as it is well regenerated under a proper cyclic or continuous regeneration mode.

Fig. 7 shows the results obtained from a four-day test in our two beds type of circulating fluidized bed reactor system at the conditions of feed composition 5%Ar/5%CO/CH4, space velocity 2500 mL/g/h, catalyst circulation rate 0.5-1.7g/min and the total amount of catalyst load 160 g. Heating the catalyst up to 1103 K or 1073 K prior to starting any second day's test was performed in H2 streams. Then it was followed by holding the catalyst at the temperature in the same streams H2 for 1-2 h to remove most of the coke accumulated in the catalyst in the previous day's test. It is worth noting here that the second day's test was conducted at 1103 K to obtain a higher benzene yield. The results demonstrates that a high CH4 conversion of >18 and a stable benzene yield of about 12% is obtainable over a considerable long period at 1073 K, but at 1103 K the catalyst will lost its stability in one day's test and more often replenishment of fresh catalyst will be needed for maintaining a stable benzene production. In comparison with the data in Fig. 6, the increased CH4 conversion and benzene yield in Fig. 7 was contributed to the presence of 5%CO in the feed. The fourth day's test was conducted in the catalyst-circulating but non-regeneration mode (N2, instead of H2, was flowed through the regenerator) and therefore the catalyst showed a rapid deactivation. The data on CH4 conversion and benzene selectivity fluctuated as the circulating rate was intentionally varied in this four days' test. This suggests that the catalyst circulation rate, which controls the mean residence times of catalyst particles in both CH4 convertor and regenerator, has a strong impact on the catalytic performance at the test conditions. Namely, optimization of the mean residence times, particularly that in CH4 convertor, is of great significance for designing and operating a real reactor system.

4. On-going work

Although the equilibrium conversion and benzene yield becomes lower at an increased pressure, a real conversion process of CH4 to benzene is very likely to be operated at a higher than atmospheric pressure to realize a possibly high process performance. Hence some pressured tests are underway. Optimization of the bed height of CH4 convertor and the velocity of feed through it are also underway.

References

1) Spivey J. J. and Hutchings G., **Chem. Sco. Rev.**, (43(2014) 792-803

2) Xu et al., **J. Catal.**, 216(2003) 386-395.

3) Honda et al., **Catal. Commun**., 4(2003)21-26

4) Cui et al., **Appl. Catal. A**, 393(2011)345-358

5) Onishi et al., **J. Catal.**, 182(1999)92-103.

6) Liu et al., **J. Catal.**, 220(2003)57-65

7)Xu et al., **Chem. Eng. Proc.,** 72(2013)90-102.

8) Xu et al., **J. Natural Gas Chem.**, 21(2012)729-744.

9) Xu et al., **Chem. Eng. J**., 168(2011)390-402.

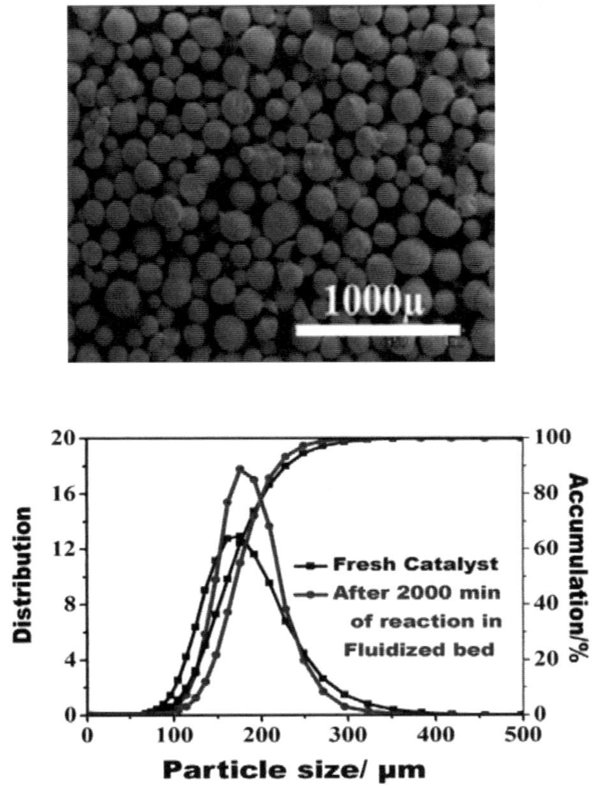

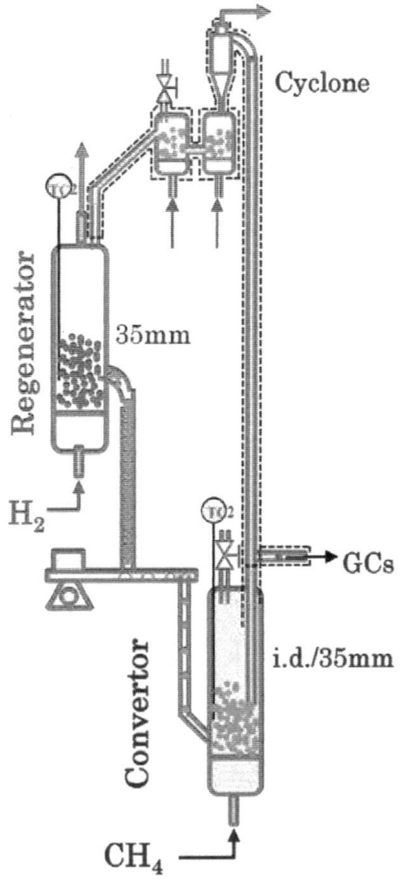

Fig. 1 A SEM image and the particle size distribution of the binder-free catalyst (left).

Fig.2 The configuration of the two beds type of circulating fluidized bed reactor system (right)

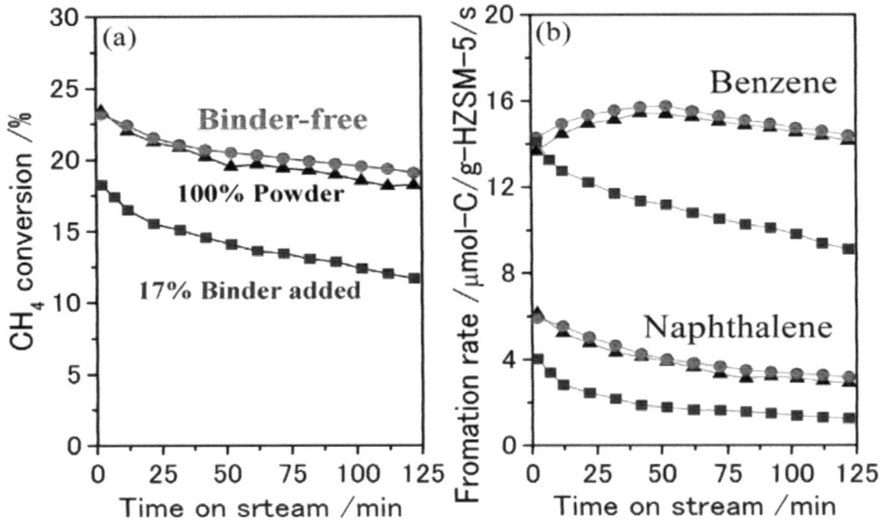

Fig. 3 Comparison of catalytic performances of the binder-free catalyst, 17% binder-added catalyst and 100%HZSM-5-based powder catalyst at 1073 K in periodic CH4-H2 switch operation mode.

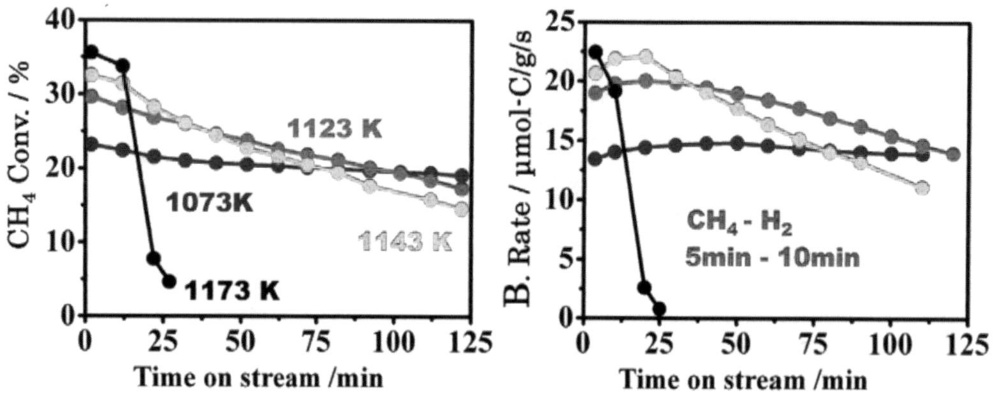

Fig. 4 Effect of reaction temperature on the activity stability of the binder-free catalyst. All tests were conducted in the micro-fixed bed reactor under periodic CH4-H2 switch operation mode.

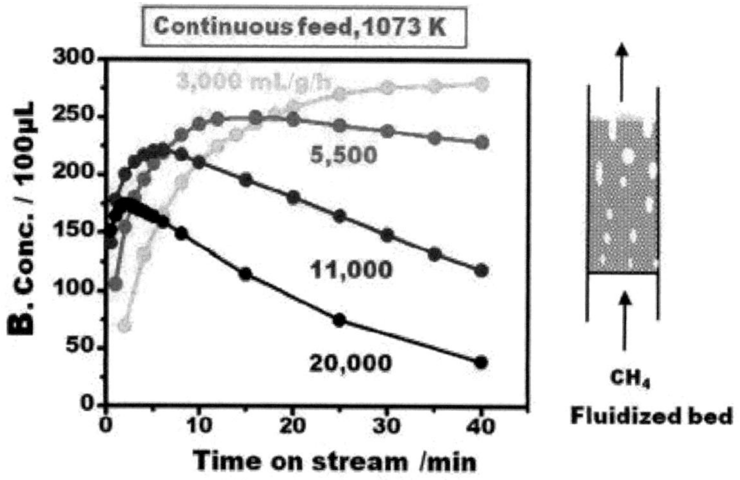

Fig. 5 Effect of space velocity on the maximum benzene formation rate of the binder-free catalyst. All tests were conducted in the micro-fluidized bed reactor in continuous feed mode at 1073 K.

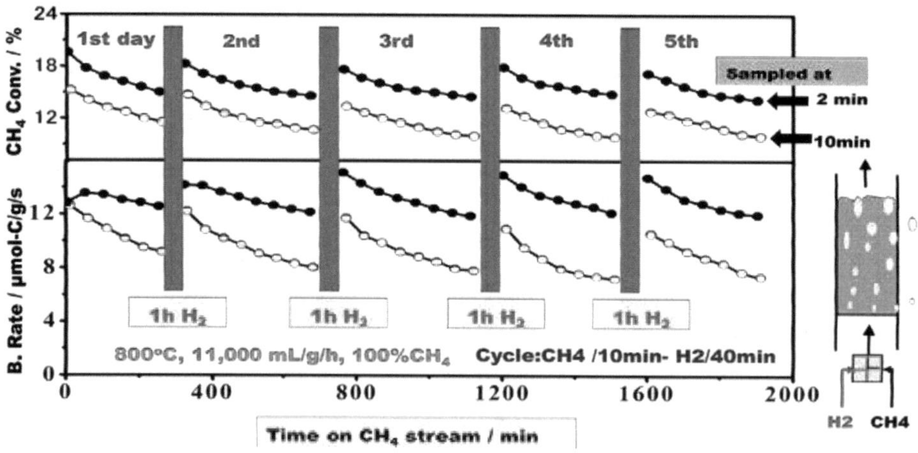

Fig. 6 Stability of the binder-free catalyst in periodic CH4-H2 switch operation mode at 1073 K.

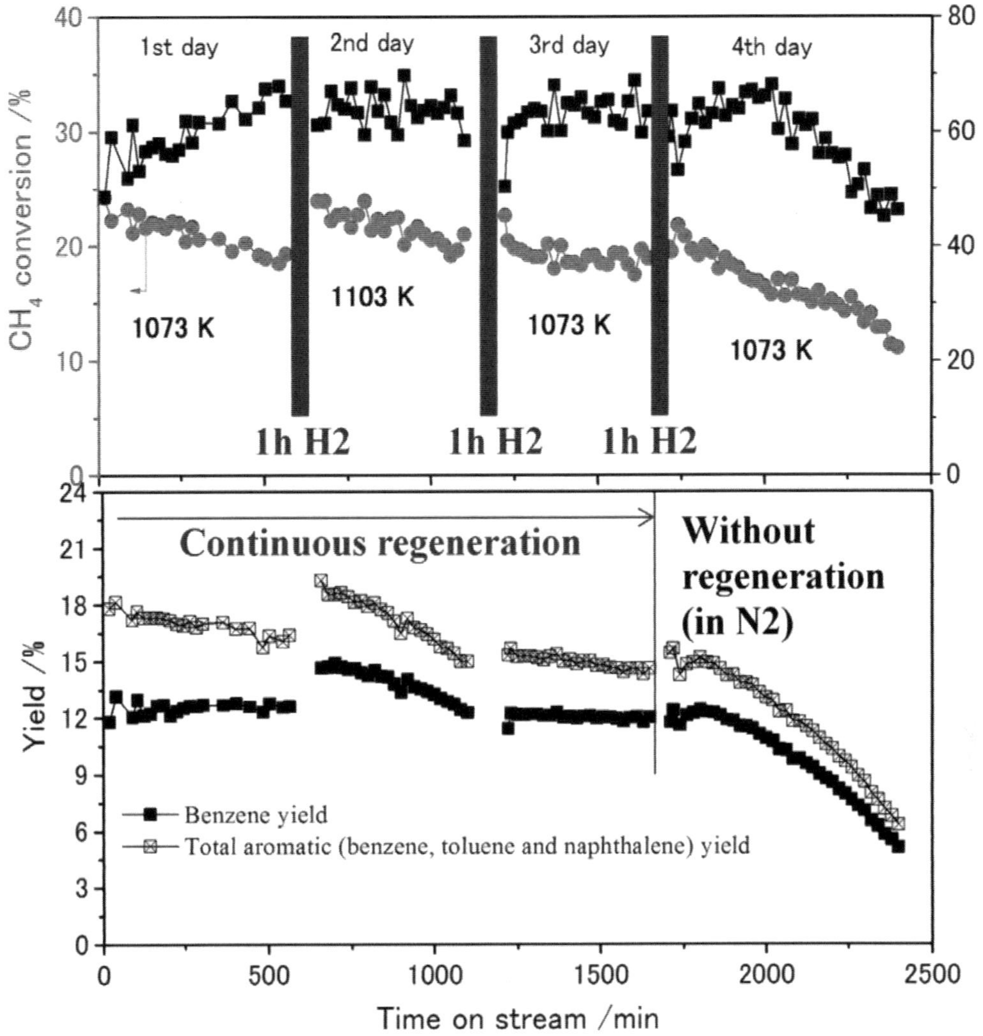

Fig. 7 Stability of the binder-free catalyst in continuous regeneration mode at 1073 K.

Chemical Looping Reforming of Methane in the Presence of Core-Shell Redox Catalysts

*Luke Neal, Arya Shafiefarhood, and Fanxing Li**

Department of Chemical and Bimolecular Engineering
North Carolina State University
Raleigh, NC 27695

* Email: fli5@ncsu.edu

Chemical looping reforming (CLR) represents a promising strategy for converting methane to syngas. In a typical CLR scheme, active lattice oxygen in a redox catalyst is used to partially oxidize methane into syngas. The oxygen depleted redox catalyst is subsequently regenerated with gaseous oxidants such as air and/or steam. For this process to be efficient, the redox catalyst needs to be selective to syngas production while maintaining high activity. Our previous work indicates that an Fe_2O_3@$La_{0.8}Sr_{0.2}FeO_3$ (LSF) core-shell redox catalyst is significantly more active and selective than traditional iron based redox catalyst. Compared with a pure LSF redox catalyst, loading of rare earth metal is significantly reduced. The present work investigates the mechanism for methane oxidation with the core-shell redox catalyst. A complex series of reaction regions is identified: (i) deep oxidation, (ii) competing selective and deep oxidation, (iii) selective oxidation with auto-activation, and (iv) methane decomposition/coke formation. The dominating mechanisms of these reaction regions are also determined through chemisorption, pulse experiments, and XPS studies. Effect of shell composition is also investigated.

392920 Overview of the U.S. DOE Fossil Energy Advanced Energy Systems Program

Abstract:

The Department of Energy's Advanced Energy Systems Program –managed within the Office of Fossil Energy and implemented by the National Energy Technology Laboratory– has been developing both second generation and transformational technologies that will drive towards developing efficient, low cost, near-zero emission fossil fuel based power conversion systems. Advanced energy conversion systems — consisting of turbines, combustion, gasification and fuel cells —are critical in reducing greenhouse gas emissions and are designed to enable low-cost, efficient carbon dioxide capture and utilization from existing and new fossil fuel power plants. The research being supported by DOE's AES Program leverages public and private partnerships to support the goal of providing multiple fossil-fuel conversion solutions to enable the continuation of low cost power generation under existing and proposed emission control regulations. Fossil Energy is targeting demonstration of 2nd generation AES technologies that result in a net efficiency gain of 4 percentage points and a CO_2 capture cost of less than $40/tonne in the 2020-2025 timeframe. Fossil Energy is also committed to exploring "step-change" transformational R&D of technologies designed to increase net efficiency gains of more than 8 percentage points with a corresponding CO_2 capture cost of less than $20/tonne by the 2035 timeframe. The AES Program is managing a portfolio of over 50 research projects from the laboratory to pilot scale (>1 Mwe). Key developments include 3,100°F natural gas and H_2 fueled turbines, supercritical CO_2 power cycles, pressurized oxycombustion, chemical looping combustion, solid oxide fuel cells and gasification. This presentation will summarize the U.S. DOE's efforts to support the development of these advanced power conversion technologies and have 2nd generation systems ready ready for commercial demonstration by 2020 and transformational systems ready for demonstration by 2035.

375681 Status of CO2 Capture Technologies

Abstract:

Fossil-powered plants are the largest point-source of CO_2 emissions and account for about 40% of all CO_2 emissions in the United States. For the past several years, The Electric Power Research Institute has critically assessed the status of CO_2 capture technologies being developed globally for post-combustion flue gas generated from such power plants. Over 130 technologies have been assessed, with a critical focus on the underlying principle of the technology, the energy consumed for the process, its potential for wide applicability, and potential need for EPRI involvement. Technologies included absorbents, adsorbents, membranes, mineralization, cryogenic, and other approaches, and spanned early-stage concepts to pilot-scale demonstrations. Summary results from this effort will be presented along with challenges of CO_2 capture, capture requirements, general status of technologies, and overall development trends.

392186 Accelerating Materials Discovery and Development for Fossil Energy Applications Using High-Throughput Computation

Abstract:

Many technological challenges associated with fossil energy applications require the development of new materials, and this task frequently requires choosing appropriate materials from a very large number of candidate materials. Significant efforts have been made to aid this kind of task by combining predictive materials modeling with high-throughput computational techniques. I will discuss examples of progress in this area and consider the challenges that exist for using computational modeling most efficiently within technology development

AIChE Topical Conference on Advanced Fossil Energy Utilization

Abstract- Plenary Session

Advanced CO2 Capture as a Source for Oil and Gas Applications

Teresa Grocela-Rocha, Technology Leader, Chemical and Environmental Systems, GE Global Research

Increasing demand for CO2 in the oil and gas industry for enhanced oil recovery is driving the need for CO2 separation from power generation and industrial sources. Advanced capture technologies which can meet acceptable cost points for these applications can serve to both increase productivity of these fossil resources while sequestering CO2. This talk will present GE's advanced aminosilicone solvent technology for CO2 capture. Targeted material selection has resulted in significant process simplification, improvements in capture cost, and reduced plant footprint. Additionally, integration of the capture plant into the source application, and optimization of the source-capture system is critical for cost effective implementation. A discussion of the CO2 capture model, optimization, and pilot validation testing will be presented.

Key Words: CO2 Capture, Aminosilicone

Mitglied der Helmholtz-Gemeinschaft

JÜLICH
FORSCHUNGSZENTRUM

A novel reactor type for autothermal reforming of diesel fuel and kerosene

AIChE Annual Meeting 2014, Atlanta, USA,
November 16 – November 21

25. November 2014 | Joachim Pasel, Remzi Can Samsun, Andreas Tschauder, Ralf Peters and Detlef Stolten

JÜLICH
FORSCHUNGSZENTRUM

HT-PEFC system with diesel and kerosene fuel processing

Target: 5 – 10 kW_e

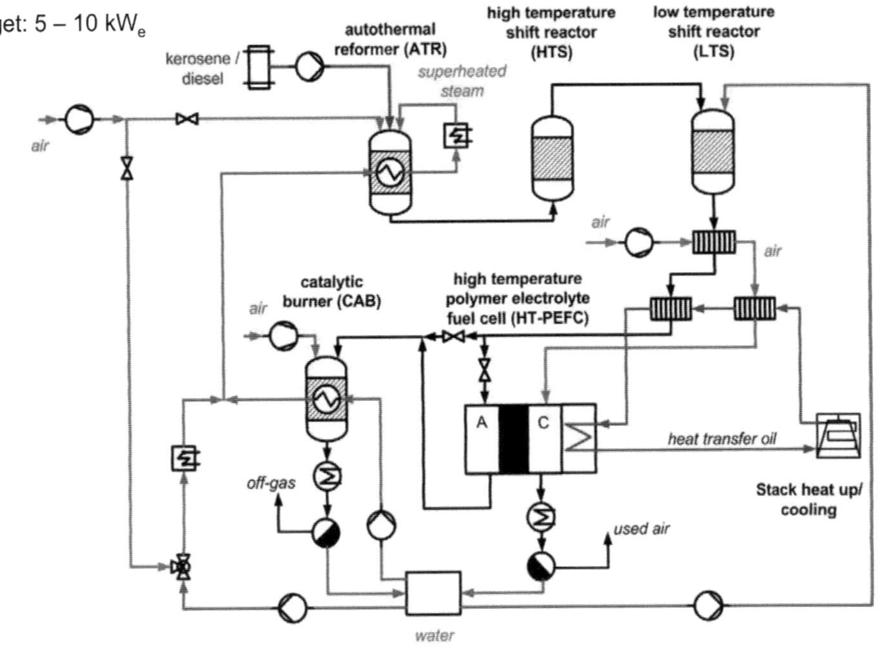

Institute of Energy and Climate Research IEK-3: Electrochemical Process Engineering

Potential applications and environmental relevance of HT-PEFC systems

Truck APU

diesel

Aircraft APU

kerosene

Sleeper trucks

3-5 kW_e

Refrigeration units

10-20 kW_e

On board power supply

100-400 kW_e

- Fuel cell systems significantly more efficient than conventional technology, especially when compared with:
  - ground operation of gas turbine APUs in aircraft or
  - truck diesel engines operating in idling mode
- Considerable savings of green house gas emissions when using fuels from bio feed stocks (palm oil, animal fat, farmed wood)

Institute of Energy and Climate Research IEK-3: Electrochemical Process Engineering

Progression in Jülich´s reactors for autothermal reforming

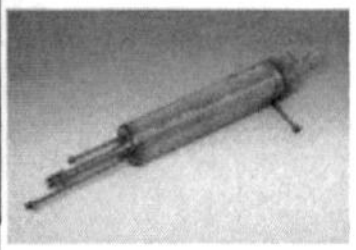

ATR 8

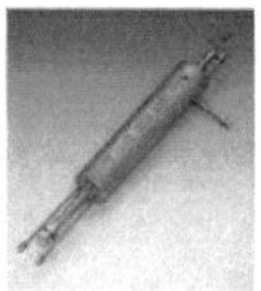

ATR 9.1

ATR 9.2

Achievements of ATR reactors with established design

- High power density and mass specific power
- Outstanding long-term stability for 10,000 h
- High quality of the product gas with carbon conversion > 99.95%

Disadvantages with respect to system integration

- External mixer (air, water, steam) and external tubing for steam supply necessary
  - More complex configuration
  - Higher heat losses
  - High pressure drop in steam generator
  - Overall efficiency reduced

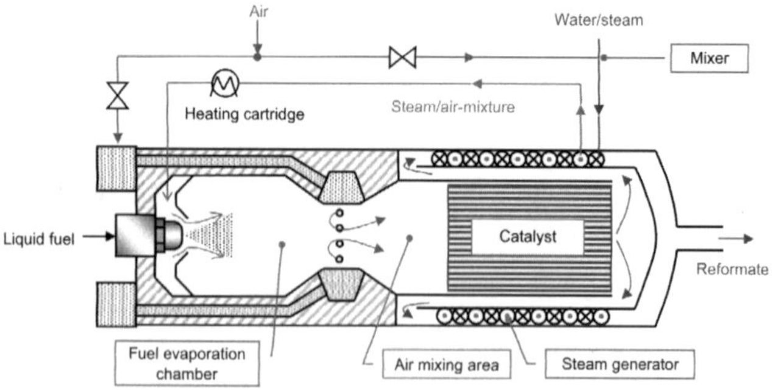

Institute of Energy and Climate Research IEK-3: Electrochemical Process Engineering

Progression in Jülich´s reactors for autothermal reforming

ATR AH2

Improvements of ATR reactors with novel design

- Incorporation of additional pressure swirl nozzle and steam generation chamber for injection of cold water
  - No need for external devices for steam supply any more
- Modification of the steam generator by increasing its cross sectional area and decreasing its length
  - Strongly reduced pressure drop in steam generator
- Heat transfer from hot reformate to saturated steam via deflecting surface and steam generator

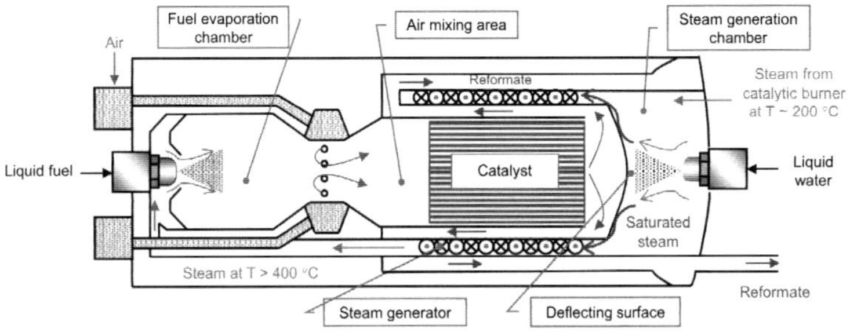

Critical temperature ranges to be kept for smooth operation of reformer and fuel processor

- Steam entering evaporation chamber: 420 °C – 600 °C
- Fuel evaporation chamber: > 360 °C
- Air mixing area: > 400 °C
- To water-gas-shift reactor: 380 °C – 430 °C

Institute of Energy and Climate Research IEK-3: Electrochemical Process Engineering

Critical temperatures at operation of novel reactor type ATR AH2
- Influence of mass fraction of cold water to nozzle

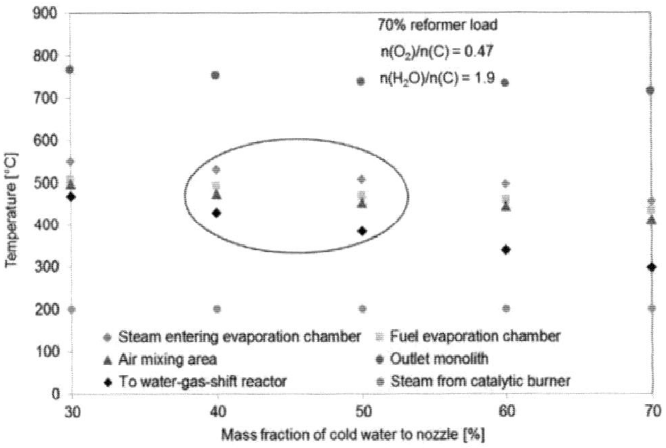

- Temperatures of steam entering the evaporation chamber, in the fuel evaporation chamber and in the air mixing area within the limits at any mass fraction of cold water to nozzle
- Temperature of reformate to water-gas-shift reactor within the limit at 40% and 50% mass fraction

Institute of Energy and Climate Research IEK-3: Electrochemical Process Engineering

Critical temperatures at operation of novel reactor type ATR AH2
- Influence of reformer load

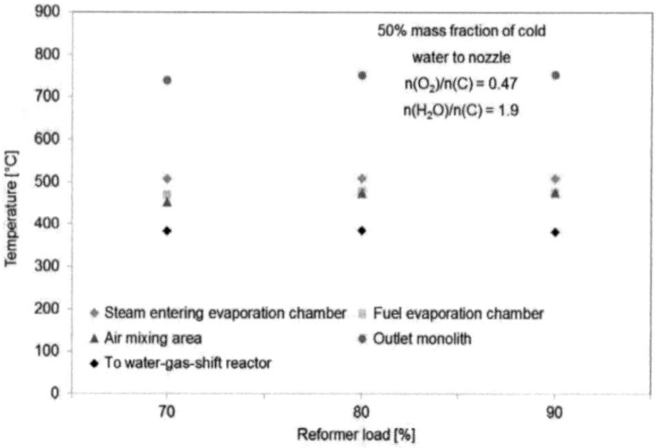

- No significant influence of the reformer load between 70% and 90% on all critical temperatures

➔ Concept for heat management of ATR AH2 proven to be applicable for fuel cell systems

Improvements of reformer operation
- Reduction of concentrations of undesired byproducts during start-up

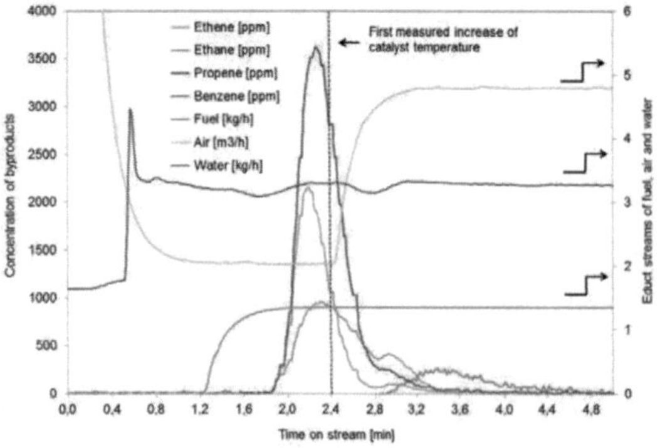

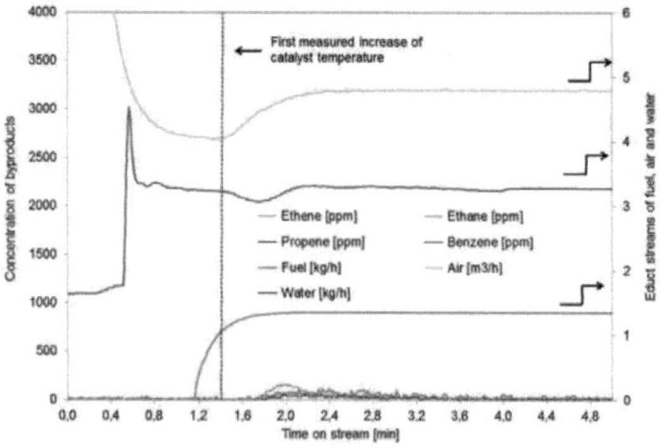

- High concentrations of ethene, ethane, propene and benzene during start-up at $n(O_2)/n(C) = 0.20$
- Increasing $n(O_2)/n(C)$ to 0.40 strongly reduced byproduct concentrations

➔ Risk of formation of carbonaceous deposits on catalysts (reformer, water-gas-shift reactor etc.) greatly reduced

Improvements of reformer operation
- Reduction of concentrations of undesired byproducts during shut-down

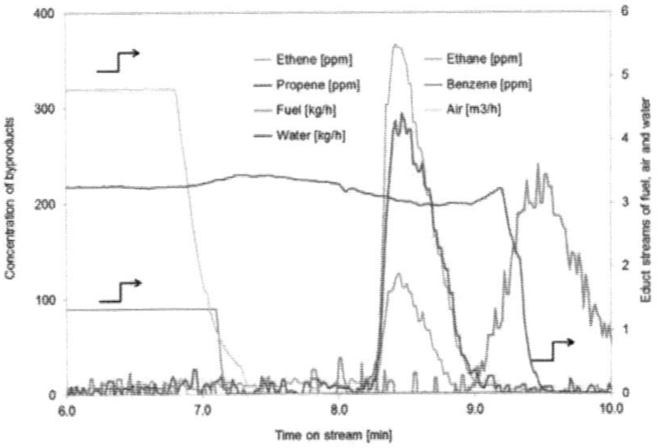

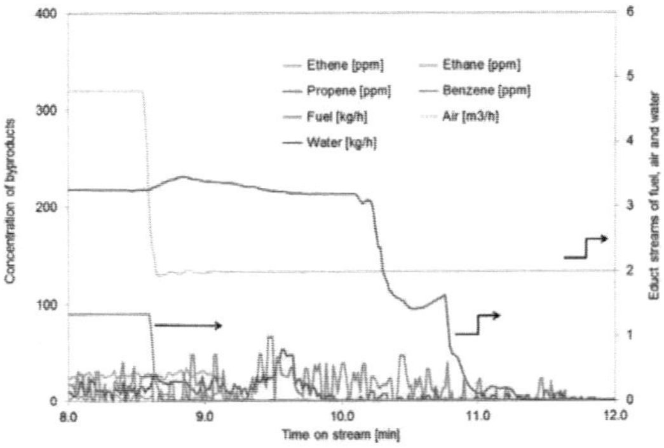

- High concentrations of ethene, ethane, propene and benzene during shut-down without air
- Retaining an air flow of 2 m^3/h strongly reduced byproduct concentrations

➔ Again, risk of formation of carbonaceous deposits on system catalysts greatly reduced

Summary and conclusions

- Novel reactor type for autothermal reforming of liquid fuels (diesel and kerosene) presented
    - Assembly of an additional nozzle and steam generation chamber for injection of cold water
    - Modification of the steam generator by increasing its cross sectional area and decreasing its length
    - Heat transfer from hot reformate to saturated steam via deflecting surface and internal heat exchanger
- Concept for heat management in novel reactor type ATR AH2 proven to be advantageous
    - All critical operation temperatures within their limits
- Significant reduction of concentrations of undesired byproducts during start-up and shut-down by increasing oxygen partial pressure
- Future experiments to evaluate the influence of improved start-up and shut-down strategies on long-term stability of reformer and further fuel cell system components

JÜLICH
FORSCHUNGSZENTRUM

Special thanks to:

Fuel Processing team at Jülich

All project and cooperation partners

Thank you very much for your attention

Parts of the work were funded by:

ADELHEID Project

Ministerium für Klimaschutz, Umwelt, Landwirtschaft, Natur- und Verbraucherschutz des Landes Nordrhein-Westfalen

progres.nrw

EUROPEAN UNION
EUROPEAN REGIONAL DEVELOPMENT FUND

Institute of Energy and Climate Research IEK-3: Electrochemical Process Engineering

391048 Pyrochlore (A2B2O7) Based Catalysts for Hydrogen Production By Hydrocarbon Reforming

Abstract:

Pyrochlores are a class of ternary metal oxides with general formula ($A_2B_2O_7$) where the A-site is occupied by the large rare-earth metal cation and the B-site is occupied by the smaller transition metal. By proper choice of the A and B-site metals, a very thermally stable crystal lattice can be obtained which can be very strong contenders as catalytic materials for high temperature applications like fuel reforming reactions. These catalysts allow isomorphic substitution of the catalytic active metals binding them to the A and B-site which results in atomic dispersion of the active metals, thus improving their resistance towards sintering at high temperatures unlike the supported commercial catalysts. The inherent lattice oxygen reactivity of pyrochlore crystal helps in increasing the pool of oxygen on the catalyst which helps in limiting deactivation due to carbon deposition. These materials have been tested for reforming of commercial diesel fuel with high sulfur and aromatic content at temperatures is excess of 900°C. Pyrochlore catalysts showed uninterrupted hydrogen production by diesel fuel reforming for a period of over 1000 hrs. Apart from aromatic and sulfur containing commercial diesel fuel reforming; these materials have also shown long lasting stable activity towards Jet-Propellant 8 (JP-8) fuel, biodiesel, and natural gas reforming for hydrogen or syngas production. Our customers have tested pyrochlore based catalysts for diesel fuel reforming in solid oxide fuel cell application and observed stable performance for about 2000 hrs. These catalysts would be of great value in high temperature reactions where other conventional catalysts would fail due to thermal instability or due to contamination of feed stream.

387261 Hydrogen Production from Renewable and Waste Oils

Abstract:

Fuel cells are a high efficiency, small footprint, low noise technologies that can be used to generate electricity for mobile and stationary power. While considerable advances have taken place in the development of fuel cells that can use hydrocarbons and other feeds (e.g. direct methanol FC), the best feedstock is still pure hydrogen. The most common technologies for hydrogen production at industrial scales are steam methane reforming (SMR), steam naphtha reforming, hydrocarbon partial oxidation (POX) and to a lesser extent gasification. In general, petroleum naphtha is the heaviest feedstock that can be used in conventional fixed bed tubular steam reformers, and even then high steam to carbon ratios are required to avoid catalyst coking. While heavy oils such as petroleum residuum, oil sand bitumen and waste oil can be gasified, gasification plants are expensive and require an air separation plant if nitrogen is to be avoided in the syngas. These heavy feedstocks cannot be steam reformed using conventional fixed bed processes.

TDA Research Inc. has developed a new technology based on chemical looping steam reforming that generates hydrogen from low cost heavy feedstocks such as vacuum residuum, oil sand bitumen, raw biomass pyrolysis oil, etc., at a cost that is considerably lower than hydrogen produced using conventional technologies (SMR, POX or gasification). In TDA's process the heavy feed is steam reformed using a nickel catalyst to produce syngas without catalyst deactivation and without the need for an oxygen plant (required when using gasifiers). Residuum and other difficult feedstocks can be steam reformed using nickel catalysts without deactivation because the system uses two fluidized beds with periodic catalyst regeneration using air. Heavy feed and steam are fed into a fluidized bed steam reformer where syngas is generated at ~870°C using a Ni catalyst. The syngas typically contains ~70 vol% H_2 (dry basis). Because the contact time in the steam reformer is on the order of minutes, not enough carbon builds up to cause irreversible catalyst deactivation. The catalyst is then regenerated in a separate fluidized bed by burning off the coke (and any sulfur) with air. In the laboratory, this is done using a single reactor with a nitrogen purge between reforming and regeneration steps; however, in an industrial setting, a circulating fluidized bed system would be used.

Burning off coke in the regenerator reheats the catalyst to ~900°C for the next reforming step (a small amount of raw feedstock can be burned in the regenerator to increase the temperature if there is not enough coke on the catalyst to generate the required heat). The hot nickel catalyst returning to the reformer is present as catalytically inactive NiO but is quickly reduced to catalytically active nickel metal by the hydrocarbons in the feed and steam reforming resumes.

TDA has conducted extensive testing using atmospheric and vacuum residuum, raw biomass pyrolysis oil and dilbit (oil sane bitumen diluted with 30% condensate to make it fluid) with no catalyst deactivation in over 600 hours of testing. Steam to carbon ratios are typically between 3 and 5, even when steam reforming petroleum residuum.

Heat-Integrated Steam Reformer for Hydrogen Production

Saurabh A. Vilekar, Christian Junaedi, Kyle Hawley, Dennis Walsh, and Subir Roychoudhury
Precision Combustion, Inc., 410 Sackett Point Rd., North Haven, CT.

Abstract

Precision Combustion, Inc (PCI) has been developing a novel heat-integrated steam reforming reactor for efficient syngas (i.e., H_2 and CO) generation based on its patented Microlith® technology, providing higher heat flux compared to other state-of-the-art reactors. The Microlith based reactor design results in a remarkably compact, lightweight, and efficient steam reforming reactor, which is otherwise widely known to be limited by the heat transfer resistance from the exothermic (heat source) to the endothermic reforming side. The burner component comprises of catalytic oxidizer instead of flame-stabilized combustion, resulting in improved temperature control and uniformity. In this paper, results from testing the reformer with a distillate fuel (containing 2-5 ppm_w sulfur) and natural gas will be presented, where near equilibrium product distribution was observed. The effect of steam-to-carbon ratio, operating pressure (i.e., up to 10 atm), and thermal inputs on the catalyst performance, fuel conversion and H_2 production will be discussed. The sulfur tolerance studies of the steam reforming catalyst and the burner catalyst will also be presented. Catalytic sulfur tolerance allows effective utilization of poor quality or waste fuels, thus significantly improving the overall process efficiency. Results from hundreds of hours of testing of the steam reformer unit showed minimal catalyst degradation, thus providing a measure of the reactor performance under realistic conditions and demonstrating the durability of the reformer. Finally, downstream hydrogen separation approaches are being examined and results will be reported. We will also discuss modeling analysis for system integration for high purity hydrogen generation.

Session: Fuel Processing for Hydrogen Production (Topical E: Advanced Fossil Energy Utilization)

AIChE 2014

November 18th, 2014

Development of diesel pre-reformer for methane rich gas production

Joongmyeon Bae[a], Sangho Lee[a], Minseok Bae[a] and Sai P. Katikaneni[b]

[a]Korea Advanced Institute of Science and Technology

[b]Saudi Aramco

Contents

1. Introduction
2. Development of diesel pre-reforming catalyst
3. Development of structured catalyst
4. Demonstration of $1kW_e$ diesel reformer
5. Conclusion

1. Introduction

2. Development of diesel pre-reforming catalyst

3. Development of structured catalyst

4. Demonstration of 1kW$_e$ diesel reformer

5. Conclusion

Solid oxide fuel cell for APU

- **Why higher APU efficiency required?**

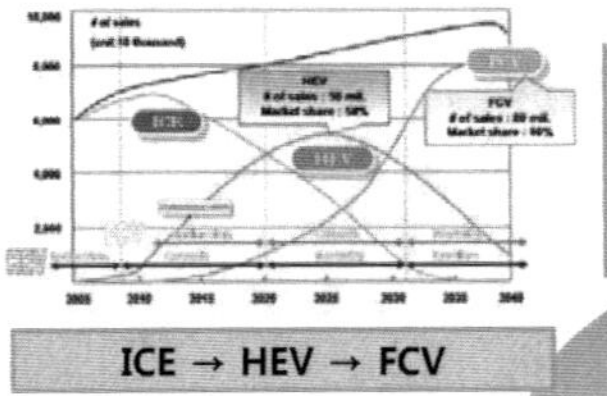

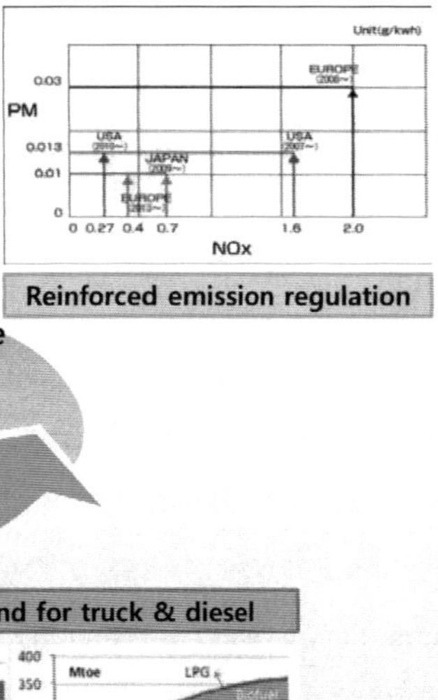

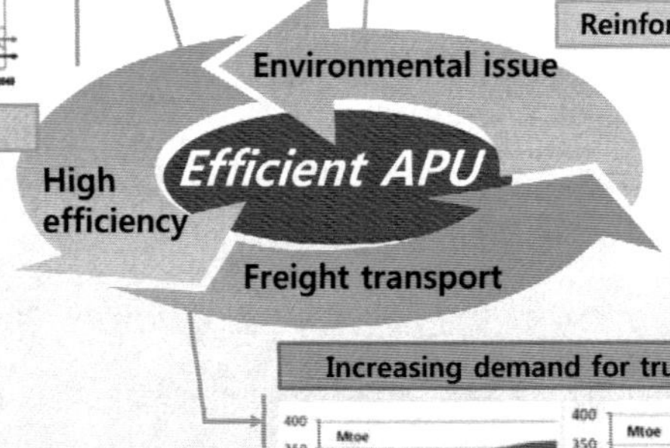

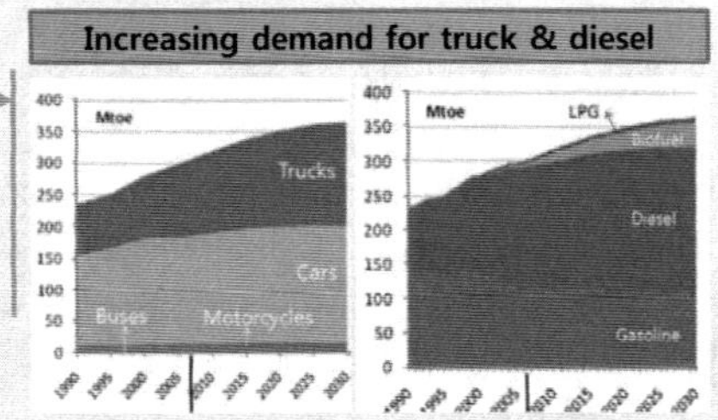

* U.S. Energy Information Administration, Annual Energy Outlook 2010-With Projections to 2035
* European Commission, European Energy and Transport (EET) – trends to 2030
* Automotive world car industry forecast report, Global insight

4

Solid oxide fuel cell for APU

- **Advantages of diesel SOFC APU**

High efficiency
- Higher efficiency than diesel generator

Low emission
- Low operating temperature for NOx formation

Compactness
- High energy density and simple configuration

Fuel flexibility
- Various fuels: Natural gas, LPG, Gasoline, Diesel, etc.

5

Solid oxide fuel cell for APU

- **Facts of Delphi diesel SOFC APU**
  - ✓ Electrical power: up to 5kW
  - ✓ <60 dB at 3 meters
  - ✓ Ultra-clean, near zero emissions
  - ✓ On-board reforming capability
  - ✓ Fuels: natural gas, diesel, etc.

6

Solid oxide fuel cell for APU

• **Projected performance of Delphi diesel SOFC APU**

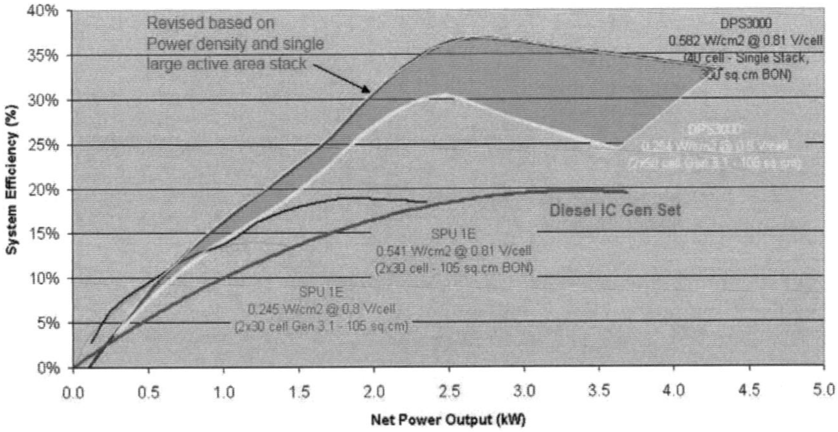

7

Solid oxide fuel cell for APU

• **Diesel reforming for SOFC APU**

Advantages

- High hydrogen density
- Broad applications
- Mature infrastructure

Issues

- Fuel delivery & Mixing
- Aromatic compounds
- Carbon deposition
- Sulfur poisoning

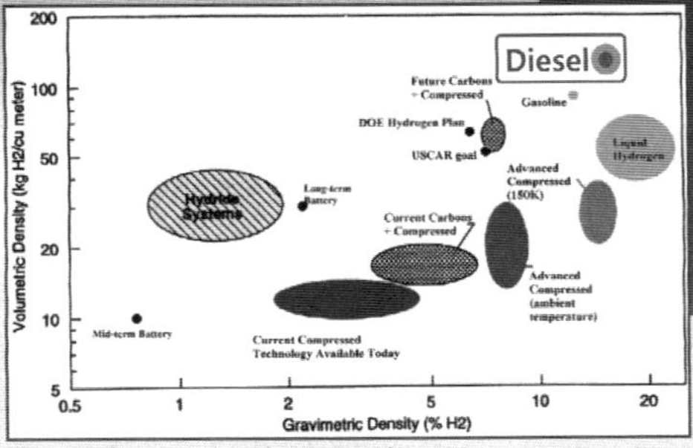

Hydrogen density to various energy sources

8

Pre-reformer for methane rich gas production

• **KAIST diesel fuel processor in phase I (2008~2010)**

KAIST diesel fuel processor

Fuel injector
- Mixing effect of reactants (fuel, air, steam)

Autothermal reforming
- Effective decomposition of Aromatics
- High selectivity of H_2 & CO
- Suppression of carbon deposition
- Sulfur tolerance of catalyst

Desulfurization
- Sulfur compounds removal for post-reforming & SOFC anode

Post-reformer
- Low-hydrocarbons (carbon precursors) removal for SOFCs

9

Pre-reformer for methane rich gas production

• **KAIST diesel fuel processor in phase I (2008~2010)**

✓ ATR reformer : 10%/2500h

✓ Post-reformer : 5%/2500h

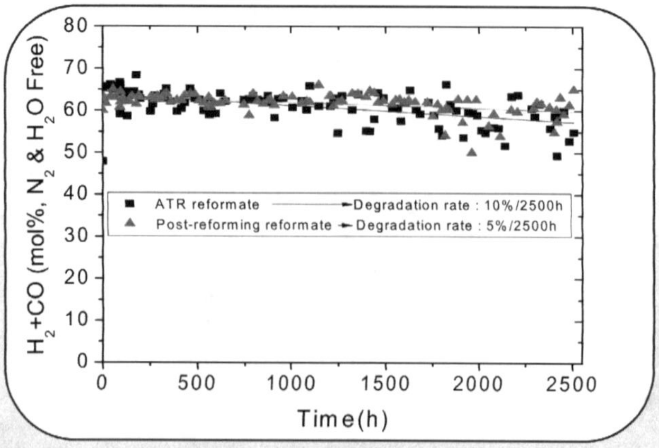

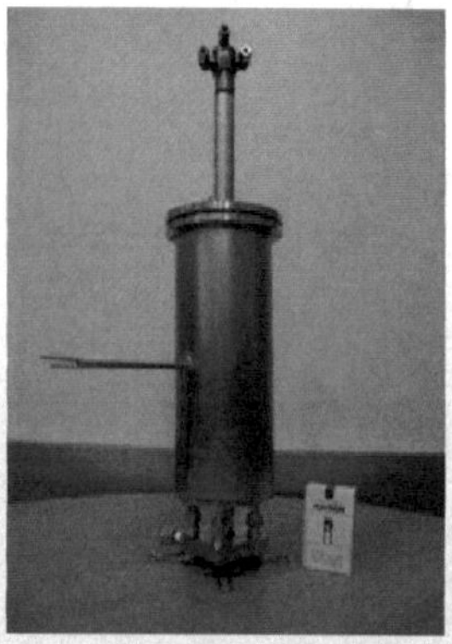

* Sangho Yoon, Sangho Lee, Joongmyeon Bae, Development of a self-sustaining kWe-class integrated diesel fuel processing system for solid oxide fuel cells, International Journal of Hydrogen Energy 36 (16), pp. 10302-10310
* Sangho Yoon, Joongmyeon Bae, Sangho Lee, Thang V. Pham, Sai P. Katikaneni, A diesel fuel processor for stable operation of solid oxide fuel cells system: II. Integrated diesel fuel processor for the operation of solid oxide fuel cells, International Journal of Hydrogen Energy, 37 (11), pp. 9228-9236
* Unified Fuel Processing Reactor for Solid Oxide Fuel Cell, 8,101,141 US patent

10

Pre-reformer for methane rich gas production

- **Problems of KAIST diesel fuel processor in phase I (2008~2010)**
  - ✓ High O_2/C ratio & Temp.
    - ➔ Increase in stability by suppressing coke formation
    - ➔ Decrease in H_2, CO and CH_4 yield

Fast degradation & Low efficiency

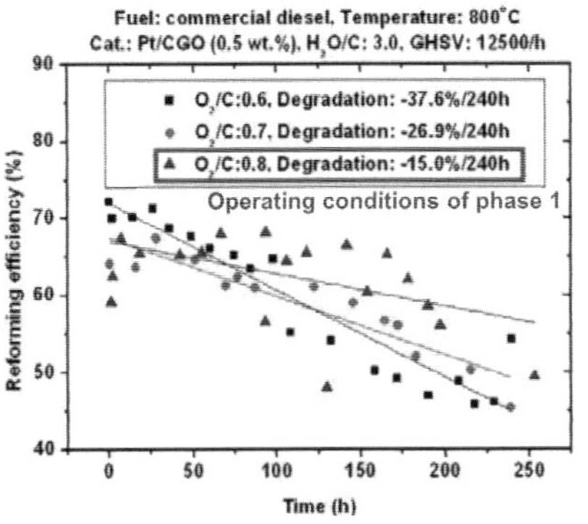

Efficiency and CH_4 vs O_2/C

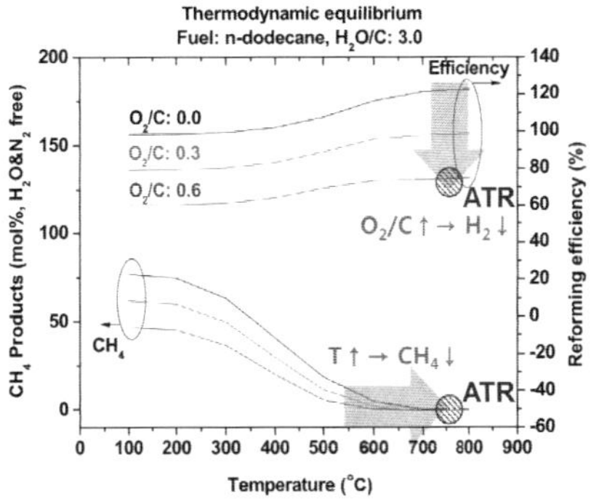

11

Pre-reformer for methane rich gas production

- **Pre-reforming conditions: $O_2/C = 0.0$, Temp. < 600°C**
  - ✓ Increase in H_2, CO and CH_4 yield

Efficiency and CH_4 vs O_2/C

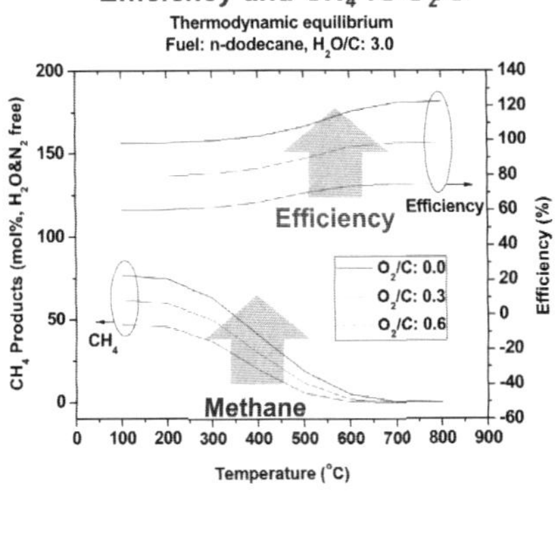

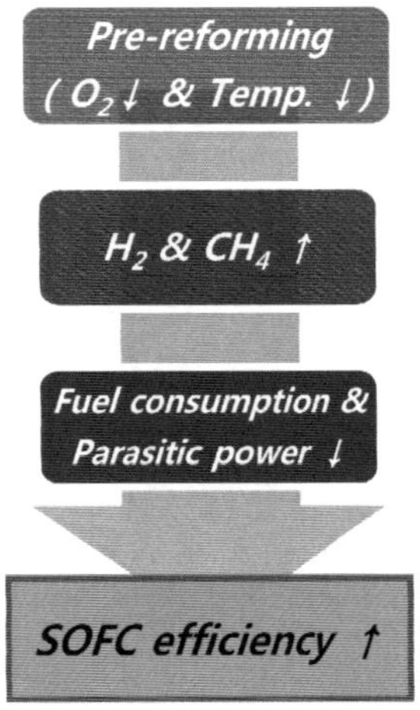

12

1. Introduction

2. Development of diesel pre-reforming catalyst

3. Development of structured catalyst

4. Demonstration of 1kW_e diesel reformer

5. Conclusion

Development of diesel pre-reforming catalysts

Catalyst development
↓
Active metal screening: Precious metals vs Ni
↓
Preparation method: Incipient wetness method vs glycine nitrate process
↓
Metal loading
↓
Promoter addition
↓
Long-term test

Catalysts candidates

- **Precious metal catalysts**

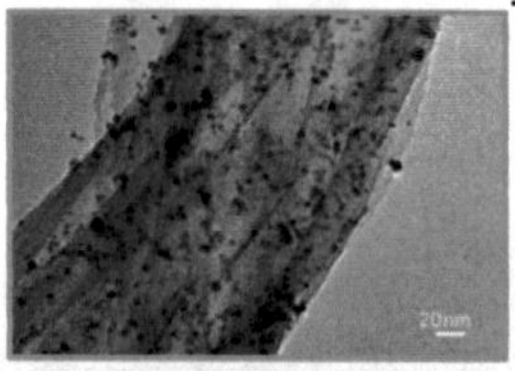

Advantages:
- High activity
- Strong to coke formation
- Strong to sulfur poisoning

Disadvantage:
- Expensive for high content

- **Ni based catalysts**

Advantages:
- Cheap for high content

Disadvantage:
- Weak to coke formation
- Weak to sulfur poisoning
- Vulnerable to oxidation

14

Development of diesel pre-reforming catalysts

- **Metal screening for diesel pre-reforming**
  - ✓ Activity for pre-reforming : Ru > Rh > Pt
  - ✓ Low activity to commercial diesel pre-reforming

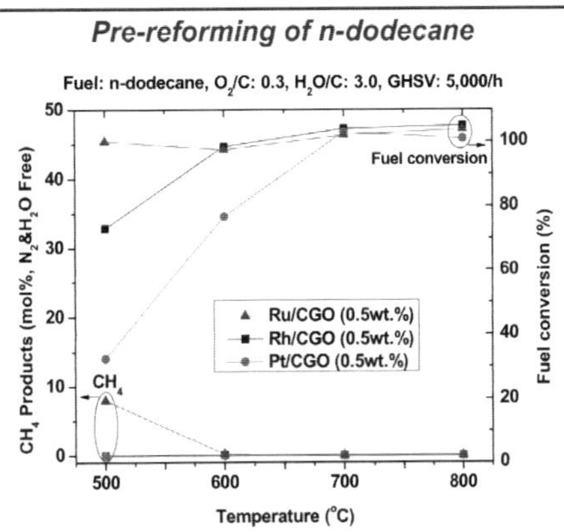

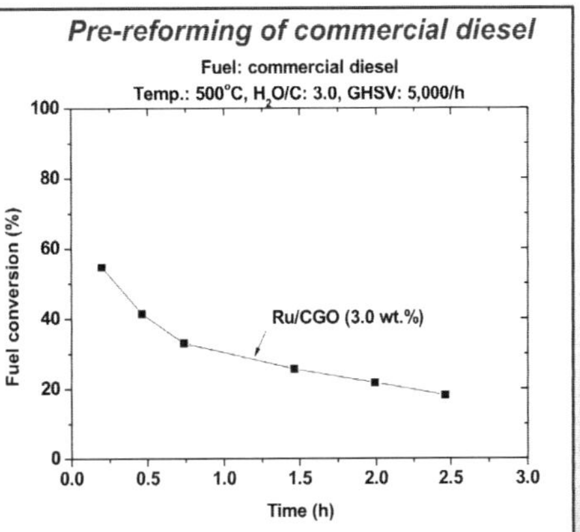

$$\text{Fuel conversion} = \frac{\text{Total number of carbon, CO, CO2 and CH4 in product}}{\text{Carbon number in fuel used}} \times 100$$

Development of diesel pre-reforming catalysts

- **Metal screening for diesel pre-reforming**
  - ✓ High activity of Ni based catalysts under 500°C
  - ✓ Degradation of 20 wt.% Ni catalyst after 4 hours

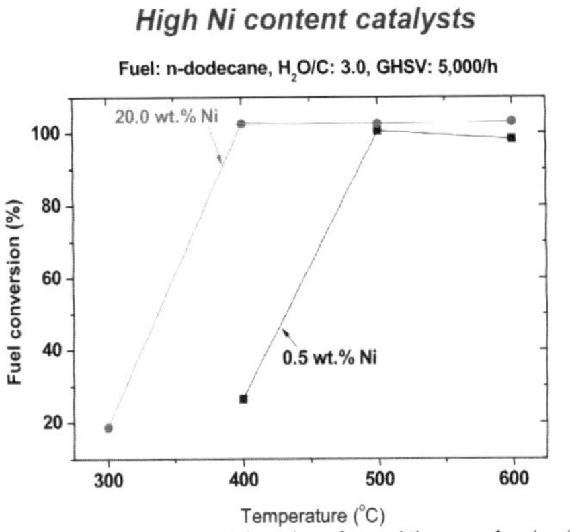

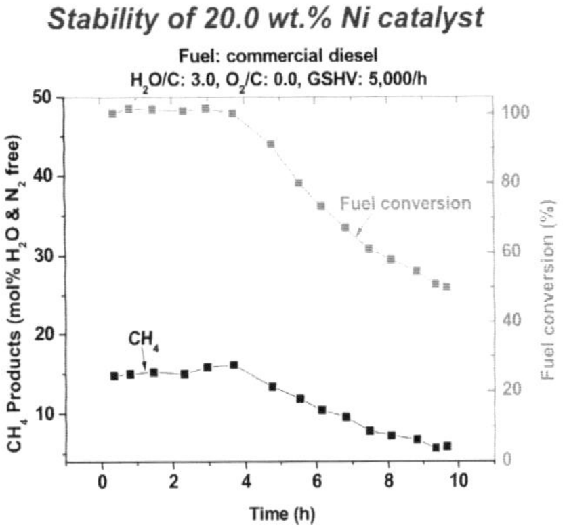

*C11PR: Ni based commercial catalysts for naphtha pre-reforming (Süd-chemie)

$$\text{Fuel conversion} = \frac{\text{Total number of carbon, CO, CO2 and CH4 in product}}{\text{Carbon number in fuel used}} \times 100$$

Development of diesel pre-reforming catalysts

- **Comparison of preparation methods**
  - ✓ Degradation rate: GNP > Incipient
  - ✓ Reduction temperature: GNP > Incipient

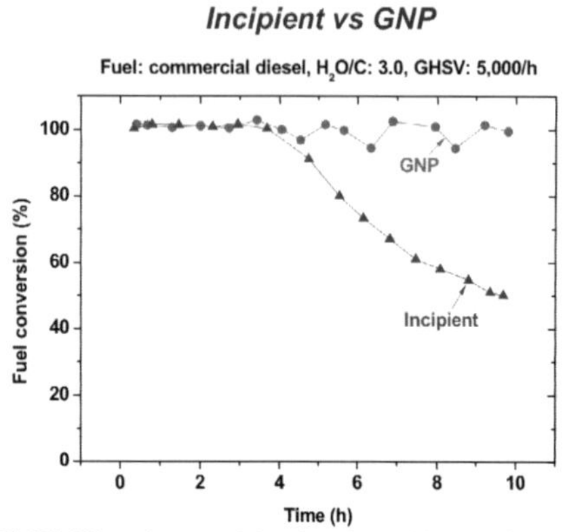

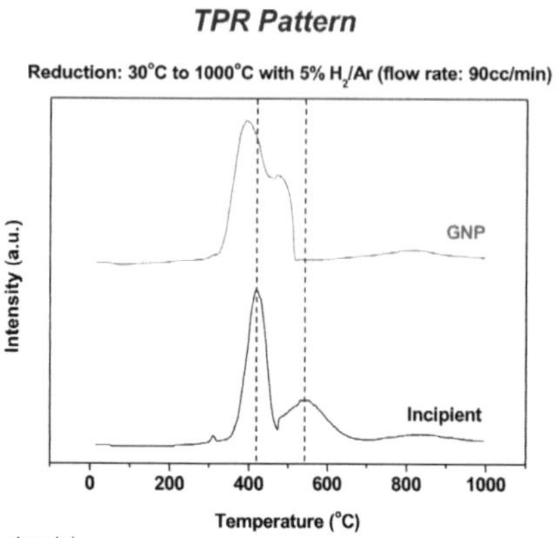

*C11PR: Ni based commercial catalysts for naphtha pre-reforming (Süd-chemie)

$$\text{Fuel conversion} = \frac{\text{Total number of carbon, CO, CO2 and CH4 in product}}{\text{Carbon number in fuel used}} \times 100$$

17

Development of diesel pre-reforming catalysts

- **Optimization of Ni contents (accelerated degradation test)**
  - ✓ Optimum Ni contents : 20 wt.%
  - ✓ Higher activity and stability than commercial catalysts

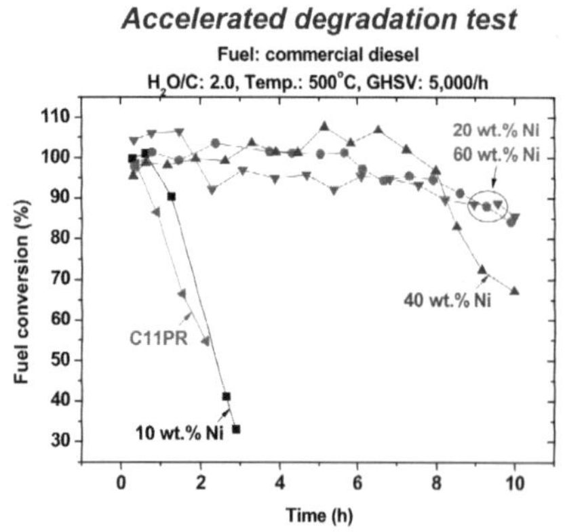

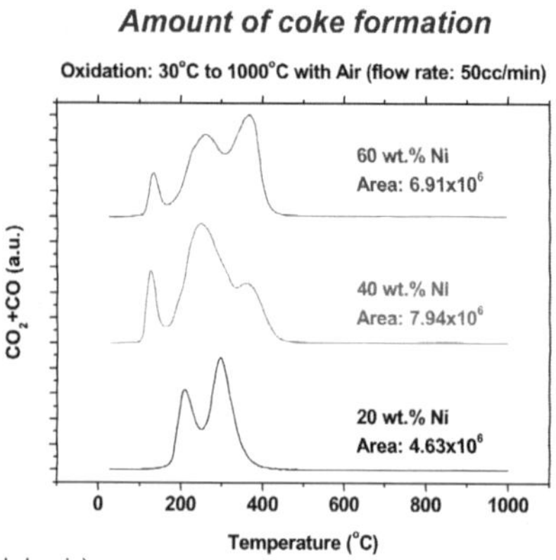

*C11PR: Ni based commercial catalysts for naphtha pre-reforming (Süd-chemie)

$$\text{Fuel conversion} = \frac{\text{Total number of carbon, CO, CO2 and CH4 in product}}{\text{Carbon number in fuel used}} \times 100$$

18

Development of diesel pre-reforming catalysts

- **Ni based catalysts with precious metals (accelerated degradation test)**
  - ✓ Improved stability by addition of Ru
  - ✓ Temperature of coke oxidation: Ni-Ru (200°C) < Ni-Pt & Ni-Rh (300°C)

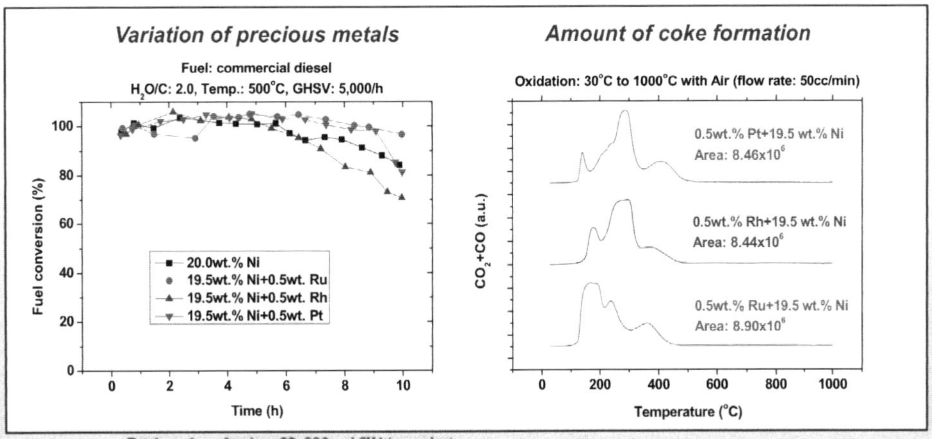

$$\text{Fuel conversion} = \frac{\text{Total number of carbon, CO, CO2 and CH4 in product}}{\text{Carbon number in fuel used}} \times 100$$

19

Development of diesel pre-reforming catalysts

- **Long-term tests of the optimized catalyst**
  - ✓ Operating time: 2,000 Hours
    - ➔ Catalyst was regenerated at 1,350 hours

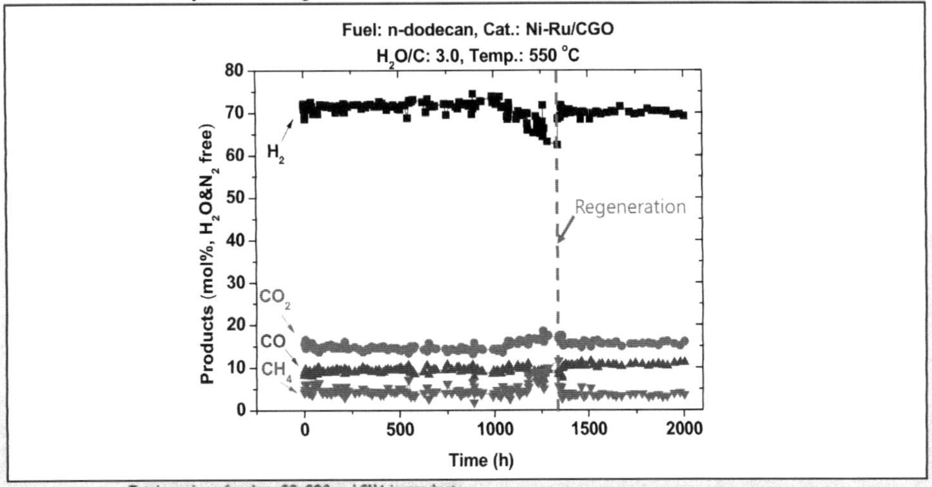

$$\text{Fuel conversion} = \frac{\text{Total number of carbon, CO, CO2 and CH4 in product}}{\text{Carbon number in fuel used}} \times 100$$

20

Development of structured catalyst

- **Preparation of structured catalyst**
  - ✓ Preparation of structured catalyst
    - ➔ CGO was pre-coated before catalyst coating
    - ➔ CGO and Ni-Ru/CGO are coated on substrates by dip-coating

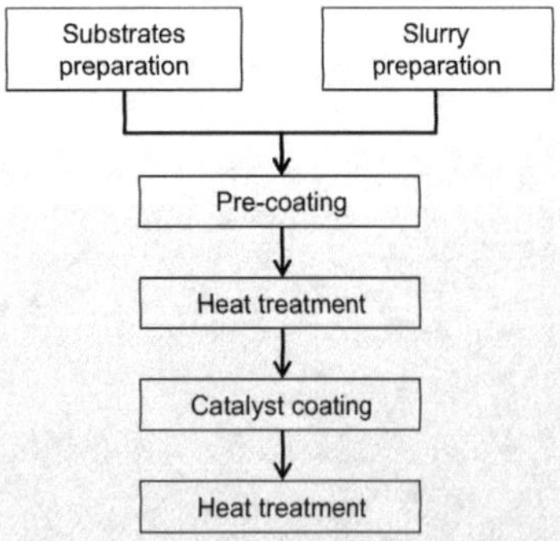

$$\text{Fuel conversion} = \frac{\text{Total number of carbon, CO, CO2 and CH4 in product}}{\text{Carbon number in fuel used}} \times 100$$

Development of structured catalyst

- **Pre-coating for structured catalysts**
  - ✓ The roles of pre-coating
    - ➔ Adhesion between catalysts and substrates can be enhanced
    - ➔ Undesired reactions (Si diffusion out, reactions on substrates and etc) can be prevented

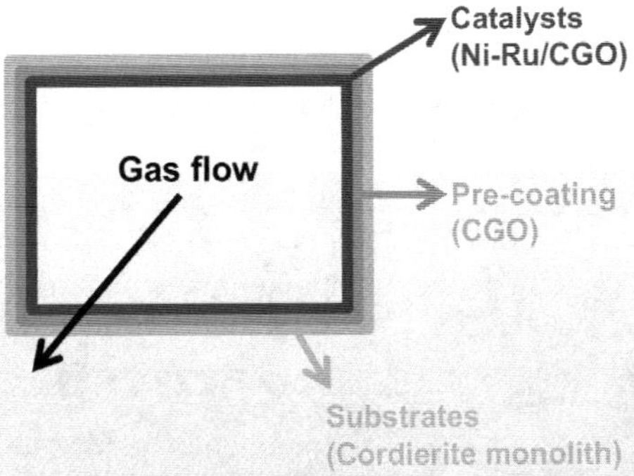

The composition of Catalyst		
Ni	19.5%	Expansion by Redox & Carbide
Ru	0.5%	
CGO	80%	

The composition of cordierite monolith		
SIO_2	45-52%	Si diffusion out
Al_2O_3	31-37%	Coke formation
MgO	5-16%	-
Fe_2O_3, Na_2O, K_2O	<3.0%	-

23

Development of structured catalyst

- **Pre-coated structured catalysts**
  - ✓ Pre-coating was introduced before Ni-Ru/CGO coating
    - ➔ The stability of structured catalyst was improved by CGO pre-coating
    - ➔ The pre-coated structured catalyst was operated for 200 hours

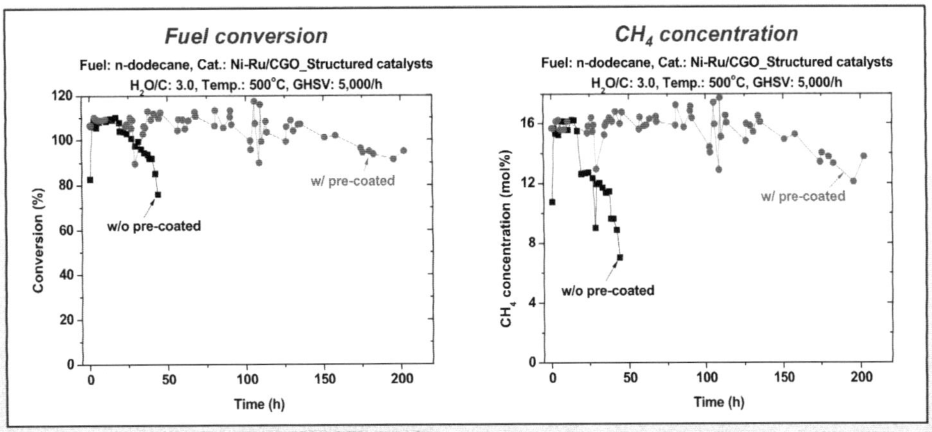

$$\text{Fuel conversion} = \frac{\text{Total number of carbon, CO, CO2 and CH4 in product}}{\text{Carbon number in fuel used}} \times 100$$

24

Development of structured catalyst

- **Morphologies of structured catalysts**

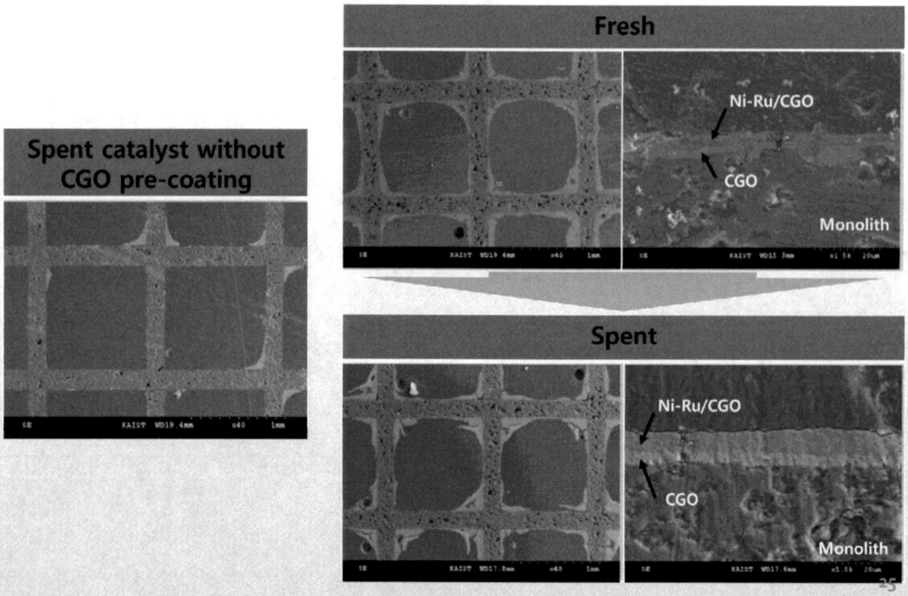

25

1. Introduction
2. Development of diesel pre-reforming catalyst
3. Development of structured catalyst
4. **Demonstration of $1kW_e$ diesel reformer**
5. Conclusion

26

Demonstration of 1kW$_e$ diesel reformer

- **Experiments preparation**

Experiment setup

Extrudate catalyst

Structured catalyst

27

Demonstration of 1kW$_e$ diesel reformer

- **1kW$_e$ diesel reformer with the extrudate catalyst**
  - ✓ The reformer was operated at 400, 600, 800 and 1,000W
    - ➔ Reformer capacity ↑ ⇒ Catalyst bed temperature ↓ ⇒ CH_4 concentration ↑

Product distribution

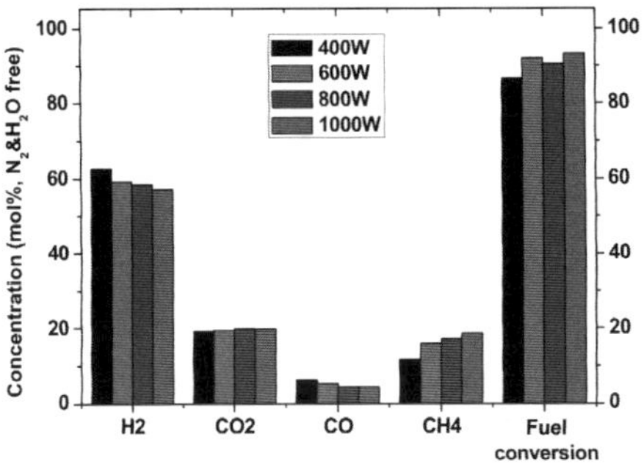

Temperature distribution

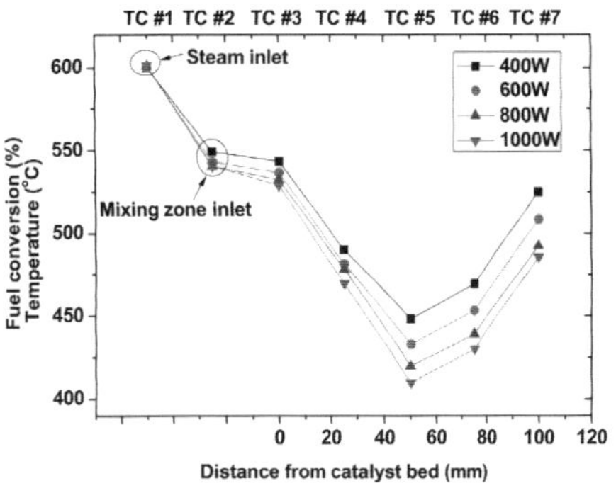

$$\text{Fuel conversion} = \frac{\text{Total number of carbon, CO, CO2 and CH4 in product}}{\text{Carbon number in fuel used}} \times 100$$

28

Demonstration of 1kW$_e$ diesel reformer

- **1kW$_e$ diesel reformer with the structured catalyst**
  - ✓ The reformer was operated with the structured catalyst
    - ➔ Reformer capacity ↑ ⇒ Fuel conversion and temperature ↓
    - ➔ Temperature decrease indicates heat transfer limitation

Product distribution *Temperature distribution*

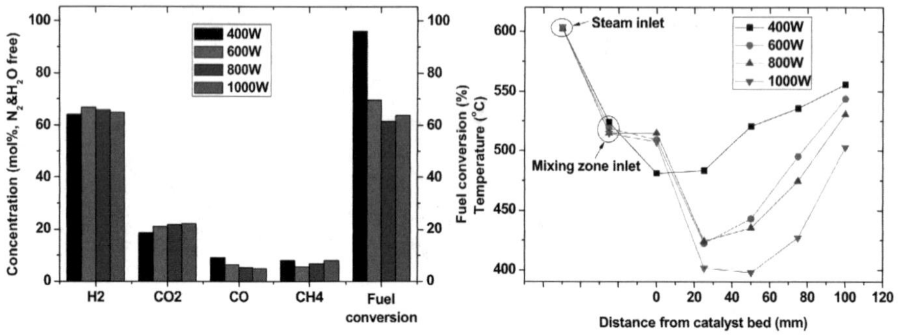

$$\text{Fuel conversion} = \frac{\text{Total number of carbon, CO, CO2 and CH4 in product}}{\text{Carbon number in fuel used}} \times 100$$

29

1. Introduction
2. Development of diesel pre-reforming catalyst
3. Development of structured catalyst
4. Demonstration of 1kW$_e$ diesel reformer
5. **Conclusion**

Conclusion

- **Diesel pre-reforming catalyst**
  - ✓ Catalyst was developed and tested for diesel pre-reforming
    - ➔ Gas composition: 61.7 mol% H_2 and 14.0 mol% CH_4
    - ➔ Catalyst durability: 2,000 hours
- **Structured catalyst**
  - ✓ Structured catalysts were prepared using the developed catalyst
  - ✓ CGO pre-coating enhanced the durability of structured catalyst
    - ➔ Gas composition: 59.2 mol% H_2 and 15.6 mol% CH_4
    - ➔ Catalyst durability: 200 hours
- **$1kW_e$ diesel reformer**
  - ✓ $1kW_e$ diesel reformer was demonstrated using the developed catalysts
  - ✓ The structured catalyst showed heat transfer limitation

31

Organization for Aramco project

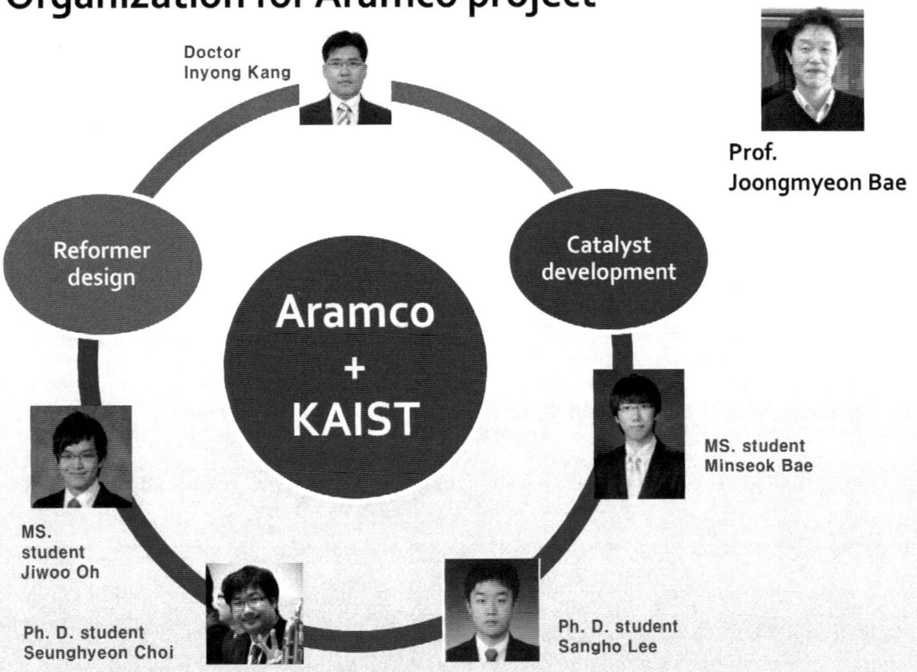

The authors wish to thank Saudi Aramco, Saudi Arabia for the financial assistance and permission to publish results

33

Desulfurization of Associated Gas for Solid Oxide Fuel Cell Applications

Feraih Sh Alenazey[a], Yousef Alyousef[a], Santosh K. Gangwal[b]
[a]Energy and Water Research Institute, King Abdulaziz City for Science and Technology, Riyadh, Saudi Arabia

[b]Southern Research Institute, Durham, NC, USA

The aim of this study is to evaluate the use of associated gas in oil fields for fuel cell applications, in particular solid oxide fuel cells (SOFC). Typical associated gas in oil field has a high H_2S content. Although rich in energy; fuel value and high pressure, the gas is simply flared and vented to avoid unsafe conditions and odor problems due to its high H_2S concentrations. In this work, desulfurization options for this gas are being evaluated using ASPEN simulation. The H_2S will need to be reduced to below the tolerance of the reforming catalyst prior to reforming and to below the tolerance of the SOFC catalyst after dilution in the reformer. Because of deep cleaning requirements for SOFC, multiple desulfurization steps are anticipated: a bulk cleaning step to remove H_2S to an intermediate level followed by a polishing step to low ppmv levels. Based on an assessment of pertinent literature and an ASPEN-based techno-economic evaluation (TEA), two process options involving various combinations of bulk and polishing have been selected for further consideration. Bulk cleaning processes that were evaluated included among others (1) amine scrubbing and (2) direct oxidation of H_2S to elemental sulfur (e.g. Crystasulf). Polishing steps evaluated included a variety of regenerable sorbents based on zinc oxide or disposable metal oxide sorbents (for example iron oxide on wood chips). Initial results of the TEA indicate that selective amine scrubbing followed by a disposable polishing sorbent to be the best option for small scale 0.1 to 1 KWe SOFC applications. The results of the TEA at various scales of operation will be presented along with associated gas-based SOFC testing and future field development and testing plans.

388842 Sulfatrap Sorbents for Effective Sulfur Removal for Fuel Cells: Dimethyl Sulfide, Mercaptan and COS Adsorption

Abstract:

The effective utilization of natural gas, liquefied petroleum gas (LPG) and biogas in fuel cell applications requires removal of odorants (sulfur compounds such as hydrogen sulfide (H_2S), dimethyl sulfide (DMS), organic disulfides, mercaptans and carbonyl sulfide (COS)) to below 0.1 ppm. Adsorption is a promising technology for accomplishing such low levels of sulfur in the feed gas to the fuel cells. TDA Research Inc. (TDA) has developed the SulfaTrap^{TM} series of sorbents that are tailored for natural gas, LPG and biogas desulfurization, and has successfully carried out multiple field demonstrations with fuel cell developers.

It is important to realize that pipeline natural gas usually has few ppm to few hundred ppm of moisture, and this has an impact on adsorptive desulfurization. Water either competes for adsorption sites or may induce side reactions depending on the odorant. Similarly, the sulfur compounds and their concentrations in LPG vary widely from supplier to supplier. Biogas contains a wide variety of sulfur species depending on the source of the biogas i.e., if the biogas is generated from a waste water treatment facility or a food processing facility or a dairy farm.

TDA's sorbent can remove both the H_2S and organic sulfur species in these gases to the ppb levels. The results from the sorbent development and field demonstrations will be presented in the meeting.

385268 Ultra-Deep Desulfurization of Hydrocarbon Fuels Using TiO2 and Ag-TiO2 Adsorbents Assisted By UV Irradiation

Abstract:

Fuel cell (FC) technologies have attracted dramatic research interests in recent years to achieve clean electric power for the "hydrogen economy". However, the development of fuel cell systems is restricted because of the demand of ultra clean fuels. Considering the protection of fuel cells, ultra-deep desulfurization of liquid fuels is an important task. Adsorptive desulfurization (ADS) is considered as a promising method for producing ultra-low sulfur fuels under ambient operation conditions. The Ag/TiO_2 and Ag/TiO_2—Al_2O_3 adsorbents developed by our research group exhibited high selectivity and sulfur capacity and can be regenerated for multiple cycles. In our previous work, the desulfurization mechanism was investigated based on the role of surface acidity and surface hydroxyl groups. In the present work, UV assisted adsorptive desulfurization technique was developed for the first time by applying TiO_2/UV system into dynamic breakthrough process. Results showed that both breakthrough and saturation capacities of TiO_2 based adsorbents were improved as observed from UV-assisted breakthrough experiments using model fuels (3500 ppmw S as benzothiophene in n-octane). The breakthrough capacity of blank TiO_2 was increased from 2.45 mgS/g to 4.05 mgS/g under UV irradiation. It is known that H_2O impurities in fuels can decrease the activity of TiO2 based adsorbents. But the sulfur removal capacities were increased by adding water as fuel additive during UV adsorptive desulfurization based on our studies. The breakthrough capacity of blank TiO_2 was further increased to 5.00 mgS/g with H_2O under UV environment. Moreover, H_2O helped maintain the activity of Ag-TiO_2 adsorbent during multiply UV adsorption-regeneration cycles. No photo-oxidation products of sulfur aromatic compounds were formed during UV-assisted adsorptive desulfurization (ADS) process, as confirmed by GC-PFPD and GC-FID.

391923 Catalytic and Technological Aspects of Reforming Methanol to Electricity inside a Fuel Cell

Abstract:

The complexity of the balance of plant of a fuel cell-fuel processor unit challenges the design and development of compact and user friendly fuel cell systems. Many of these drawbacks could be alleviated by incorporation of a methanol reforming catalyst into the anodic compartment (bi-functional anode) of a high-temperature, polymer electrolyte fuel cell (HT-PEMFC), so that methanol reforming takes place inside the fuel cell stack (internal reforming). The development of an internal reforming methanol fuel cell (IRMFC) poses an ambitious technological and research challenge, which requires the effective combination of various technological approaches as regards materials development, chemical reaction engineering and stack design. It aims at opening new scientific and engineering prospects, which may allow easier penetration of the fuel cell system in the energy market. This paper particularly emphasizes catalytic and technological aspects of the methanol reformer and the IRMFC unit and reports on recent progress from laboratory prototypes.

391154 A 5 Kwel Propylene Glycol Microchannel Fuel Processor As Hydrogen Source for Fuel Cells

Abstract:

The reforming of biomass-derived oxygenates such as ethanol or glycerol has been deeply studied. However, the stability of catalyst formulations developed for glycerol reforming was not the main focus of this development work to date. In the case of propylene glycol, few studies have been addressed to date. Propylene glycol serves as coolant in many mobile applications nowadays and therefore a distribution grid already exists.

The practical processing of larger amounts of polyalcohols such as propylene glycol to synthesis gas for fuel cell applications in reactors larger than laboratory scale suited for distributed applications has not been adequately addressed to date.

The current work deals with (a) the development of stable catalyst formulations for propylene glycol reforming and (b) reformer and fuel processor development for the steam reforming of propylene glycol in the power range exceeding 17 $kW_{thermal}$ of the synthesis gas product.

A self developed stable catalyst formulation was incorporated into a plate heat-exchanger reformer. The energy required to drive the endothermic steam reforming reaction was foreseen to originate from the combustion of fuel cell anode off-gas and additional propylene glycol in a future technical process. Therefore the demonstrator was composed of layers for energy supply carrying a micro-structured catalytic burner, which was fed with anode off-gas surrogate and reaction layers which carried catalyst for the oxidative steam reforming of propylene glycol.

A complete integrated and compact fuel processor is under development including start-up burner and CO clean-up. It is to the authors' knowledge the first fuel processor for steam reforming of polyalcohols of that size presented in open literature.

Biogas steam and oxy-steam reforming reactions over Me/CeO_2–based (Me = Rh, Pt, Ni) structured cordierite monoliths

Antonio Vita[1], Giuseppe Cristiano[2], Cristina Italiano[1], Lidia Pino[1], Stefania Specchia[2]

[1] *CNR-ITAE "Nicola Giordano", Via Salita S. Lucia sopra Contesse 5, 98126 Messina – Italy*

[2] *Dept. of Applied Science and Technology, Politecnico di Torino, Torino (Italy)*

Increasing of energy requirements, dependence on crude oil, limited fossil resources, and environmental problems are huge challenges for our societies. Within such a context, fuels derived from renewable resources (biomass and waste) could give an important contribute to meet current and future energy requirements. Biogas is an attractive renewable energy source, which does not contribute to carbon dioxide emissions [1]. In recent years, some specific legislative tools of European Union (EU) have been aiming to increase the integration of renewable resources into the EU energy system [2]. Thus, biogas can be considered as the most widespread renewable fuel obtained from biomass, produced from a variety of organic raw materials ranging from zoo-technical to agro-industrial sectors.

Biogas composition is related to the starting substrate [3]. Basically it is constituted by 50-75% CH_4, 25-45% CO_2, 2-7% H_2O (at 20-40 °C), 2% N_2, less than 1% H_2/H_2S and traces of O_2, NH_3, halides and siloxanes [4]. Internal combustion engines and gas turbines represent the most employed technologies for energy generation from biogas, but at the same time show low electrical conversion (18-25%, especially in the range of 5–100 kW_{el}), high levels of noise and harmful environmental emissions. However, an alternative and sustainable use of biogas is feasible and it is based on the production of pure methane by purification processes [5] or syngas by reforming processes [6].

Biogas reforming is essentially CO_2 reforming of CH_4, even known as dry reforming of CH_4:

DR: $CH_4 + CO_2 \rightarrow 2CO + 2H_2$ $\Delta H°_{298} = 247.3$ kJ mol^{-1}

One of the major drawbacks of DR is the requirement of high temperatures to reach high conversion levels. Moreover, there is a significant potential for carbon deposition over the surface of the catalyst. Severe operating conditions, in fact, result in deactivation by coke deposition and/or sintering of the metallic phase and support. Appropriate catalysts for biogas reforming reactions must able to efficiently convert CH_4 and CO_2 in presence of H_2O for SR and H_2O and O_2 for OSR. Ni or precious metals, such as Ru, Rh, Pd, and Pt, are reported to be active as the catalyst for the mentioned reaction; however, the reactions are frequently accompanied by coke formation, especially on Ni catalysts, leading to catalyst deactivation or plugging of the reactor. Noble metal shows high selectivity for coke-free operation, which can be ascribed to high reforming rates combined with low coke formation rates [7] however, they are more expensive. A way to improve the anti-coking property of catalyst is to use supports with strong interaction Me-support (CeO_2, ZrO_2, TiO_2) coupled with high oxygen storage properties [8,9]. Heavier rare earth metal oxides, such as ceria, enhance coke gasification dramatically while maintaining or only slightly increasing steam reforming activity supposedly caused by strong metal-support interaction between metal and ceria [10,11]. Moreover, process intensification is required to achieve good performance targets especially for small-scale applications such as for distributed hydrogen production.

In this work, the performances of structured catalysts, based on Rh, Pt, Ni (noble metals load equal to 1.5 wt.%, Ni load equal to 7.5 wt.%) supported on CeO_2, coated on cordierite monoliths (400 cpsi, diameter 1 cm, length 1.5 cm) were investigated. Monoliths were lined by combining the Solution Combustion Synthesis with the Impregnation technique [12,13]. The layers are based on Me (Me = Rh, Pt, Ni, Noble Metals load = 1.5 wt.%, Ni load = 7.5 wt.% respect the CeO_2 carrier) supported on ceria oxide (CeO_2). The deposition process was repeated several times until reaching the desired amount of catalytic layer over the bare monolith (35.4 wt.%). The monoliths were finally calcined at 400 °C in calm air for 2 hrs, reaching at the end of the lining process total loads of 6.5 mg cm^{-2} referred as weight of active metal plus oxide carrier on the bare monolith geometric surface [14]. In addition to the preparation of the coated monoliths, the corresponding catalysts at powder level were prepared and used for the chemical-physical characterization of the supported catalytic materials. All prepared catalysts were characterized by XRD, TEM, TPR, BET, chemisorption analyses. The pressure drop, at different superficial velocities, across the bare cordierite and the prepared catalyzed monoliths was determined. The adherence between the catalytic layer and the cordierite walls was evaluated via sonication determining the weight loss after exposure of the catalytic monoliths to isopropyl alcohol ultrasonic treatment.

Biogas OSR and SR experiments were conducted in a fixed-bed quartz reactor with inner diameter of 1 cm at atmospheric pressure. The structured catalysts were placed between two quartz wool plugs in the center of the quartz tube, inserted into a furnace heated at the reaction temperature T_{SET} and controlled through a PID temperature controller. In a preliminary investigation the effect of S/C and both S/C and O/C (O = molecular oxygen) was valued on the Rh based structured catalyst varying the S/C (1–5) at fixed temperature (T_{SET} = 900 °C) and weight space velocity (WSV = 72,000 Nml g_{cat}^{-1} h^{-1}) for the SR reaction, and O/C (0.05–0.2) and S/C (0.3–1) at fixed temperature (800 °C) and same WSV for the OSR reaction. Instead, the influence of temperature and of the WSV was assessed on all the prepared structured catalysts varying the temperature (T_{SET}) between 700–900 °C and the WSV between 34,000 and 19,600,000 Nml g_{cat}^{-1} h^{-1} for the SR reaction, and the temperature (T_{SET}) between 750–900 °C and the WSV between 36,000 and 25,4000,00 Nml g_{cat}^{-1} h^{-1} for the OSR reaction. Model biogas containing 60% CH_4 and 40% CO_2, was prepared from high purity CH_4, O_2 and N_2 gas cylinders. Steam was added to the feed by using an isocratic pump and a specially designed evaporator.

The feed and reaction products were analyzed using an Agilent 6890 Plus gas chromatograph equipped with thermal conductivity (TCD) and flame ionization (FID) detectors. Before testing the catalytic performance of the prepared monoliths, each coated monolith was reduced under hydrogen-rich atmosphere directly in the quartz reactor of the test rig.

Catalyzed monoliths showed a uniform thin coating with thickness between 20–25 μm, high mechanical strength and low pressure drop. Regarding the catalytic activity, at low space velocity all of the catalysts showed similar performances. Instead, increasing the space velocity, the Rh/CeO_2 monolith performed better than the other samples for both SR and OSR processes. A comparison between the reforming processes efficiency based on the energy balance (available energy of outlet syngas on inlet required energy) for each process was addressed.

[1] Mollersten K. Yan J., Moreira J. R. Potential market niches for biomass energy with CO_2 capture and Storage: opportunities or energy supply with negative CO_2 emissions. Biomass Bioenergy 25 (2005) 273–285.

[2] Specchia S. Fuel processing activities at European level: A panoramic overview. Int. J. Hydrogen Energy in press (2014) doi: 10.1016/j.ijhydene.2014.04.040

[3] Kolbitsch P., Pfeifer C., Hofbauer H. Catalytic steam reforming of model biogas. Fuel 87 (2008) 701–706.

[4] Edelmann W. Biogas production and usage. In: Kaltschmitt M, Hartmann H, editors. Energy from biomass: basic principles, technologies and processes. Leipzig, Germany: Springer; 2001.

[5] Lau C.S., Tsolakis A., Wyszynski M.L. Biogas upgrade to syn-gas (H_2-CO) via dry and oxidative reforming. Int. J. Hydrogen Energy 36 (2011) 397–404.

[6] Montanari T., Finocchio E., Salvatore E., Garuti G., Giordano A., Pistarino C., Busca G. CO_2 separation and landfill biogas upgrading: a comparison of 4A and 13X zeolite adsorbents. Energy 36 (2011) 314–319.

[7] Nematollahi B., Rezaei M., Khajenoori M. Combined dry reforming and partial oxidation of methane to synthesis gas on noble metal catalysts. Int. J. Hydrogen Energy 36 (2011) 2969–2978.

[8] Ashrafi M., Proll T., Pfeifer C., Hofbauer H. Experimental study of model biogas catalytic steam reforming: 1. thermodynamic optimization. Energy Fuels 22 (2008) 4182–4189.

[9] Pino L., Vita A.S., Cipitì F., Laganà M., Recupero V. Catalytic performance of $Ce_{1-x}Ni_xO_2$ catalysts for propane oxidative steam reforming. Catal Lett 122 (2008) 121–130.

[10] Zhang Q., Qin Y., Chang L. Promoting effect of cerium oxide in supported nickel catalyst for hydrocarbon steam-reforming. Appl. Catal. 70 (1991) 1–8.

[11] Wei Y., Wang H., Li K., Zhu X., Du Y. Preparation and characterization of $Ce_{1-x}Ni_xO_2$ as oxygen carrier for selective oxidation methane to syngas in absence of gaseous oxygen. J Rare Earths 2010;28(1):357-361.

[12] Specchia S., Finocchio E., Busca G., Specchia V. Combustion synthesis, Lackner M, Winter F and Agarwal AK, Handobook of combustion, Wiley-VCH Verlag GmbH & Co. KGaA, Volume 5, Chapter 17, Weinheim (Germany), 2011, 439–472.

[13] Amjad U., Vita A., Galletti C., Pino L., Specchia S. Comparative study on steam and oxidative steam reforming of methane with noble metal catalysts. Ind. Eng. Chem. Res. 52 (2013) 15428–15436.

[14] Vita A., Cristiano G., Italiano C., Specchia S., Cipitì F, Specchia V. Methane oxy-steam reforming reaction: performances of $Ru/\gamma-Al_2O_3$ catalysts loaded on structured cordierite monoliths. Int. J. Hydrogen Energy in press (2014) doi: 10.1016/j.ijhydene.2014.03.114

Symposium on "Fuel Processing for Hydrogen Production"
Topical E: Advanced Energy Utilization
2014 AIChE Annual Meeting
Atlanta, GA
November 16-21, 2014

Production of High-Purity Hydrogen from Glycerol for Fuel Cell Applications

Devinder Mahajan[1,2], Hazem Tawfik,[3] Yue Hung,[3] Scott Turn[4], Ofei Mante[2]

[1]Materials Science & Engineering Department, Stony Brook University, Stony Brook, New York, USA
[2]Sustainable Energy Technologies Department, Brookhaven National Laboratory, Upton, New York, USA
[3]Farmingdale State College, Farmingdale, New York, USA
[4]Hawaii Natural Energy Institute, University of Hawaii at Manoa, Hawaii, USA

ABSTRACT

Two Proton Exchange (PEM) fuel cells have been installed at Farmingdale State College. The first fuel cell is a 5 kW commercial Horizon graphite air-cooled power stack. This system is selected for its simplicity, efficiency, and features a high power density, thus opening new possibilities for testing and integration with inverters and converters to power home appliances and electric vehicles. The system utilizes 75 SLPM H_2 of 99.999% purity, with maximum open circuit voltage of 100 volts DC and its operating condition are 72 V and 70 amps. The second fuel cell is a 1 kW stainless steel air cooled 25 bipolar plates with active area of 10x10 inches. These Combined Heat and Power (CHP) systems are now serving as test beds for optimization and integration with a H_2 production system for complete system demonstration.

Proton Exchange Membrane (PEM) fuel cells operate at low temperatures (< 90°C) and utilize H_2 to produce stationary power or power a vehicle with efficiencies that can be double that of an internal combustion engine. Therefore, the production of carbon-free fuel hydrogen (H_2) from renewable sources is of global interest. There are two key requirements to make H_2 fuel truly sustainable for PEM application: 1) the feedstock for H_2 production should be a renewable source and 2) stringent H_2 purity requirements due to Pt or other catalyst used in PEM fuel cells. The U.S. production of biodiesel was 950 million gallons (~16% of global production) (Biodiesel Magazine, April 2014). During biodiesel production, the plant also produces 10% by weight glycerol as a by-product and this is presently not utilized. We are, therefore, looking at the potential of H_2 from glycerol. Glycerol is attractive for its high H_2 content: glycerol contains oxygen, it has the ability to produce syngas with theoretical H_2/CO mole ratio of 4/1 by direct decomposition (Equation 1) and 7/1 by the steam reforming reaction (Equation 2).

$$C_3H_8O_3 \rightarrow 4\,H_2 + 3\,CO \qquad (1)$$

$C_3H_8O_3 + 3\ H_2O \rightarrow 7\ H_2 + 3\ CO_2$ (2)

Preliminary results of H_2 production from glycerol catalyzed reformation were performed in a 300 mL Parr batch reactor. With glycerol diluted with 10% water and after heating at < 200°C in the presence of a Mo catalyst under 100 psi N_2 gas, noted gaseous components analyzed by gas chromatography showed H_2 ~ 2%. However, the presence of other gases, CO, CO_2 and CH_4, show that there is potential to increase the yield of H_2 to a theoretical yield of 7 moles/mol glycerol.

Further integration of a subsequent step to drive water-gas-shift (WGS) reaction to reduce CO below a few ppm would produce a system that is capable of delivering H_2 of the required purity (>99.999%) for feeding directly in to a fuel cell. For the WGS step, we are exploring a Ru catalyst that can effectively operate in homogeneous mode and almost completely convert CO to CO_2 in the liquid phase at T < 200°C (Equation 3), thus avoiding the thermodynamic limitation of the gas-phase WGS reaction.

$CO + H_2O \rightarrow H_2 + CO_2$ (3)

The batch mode catalyst performance results, along with catalyst characterization studies, will be discussed in the presentation.

*This work was supported by the National Science Foundation under the Center for BioEnergy Research and Development (CBERD) under the Industry/University Cooperative Research and Development (I/UCRC) program and a New York State Energy Research and Development Authority (NYSERDA) at Stony Brook University. The work was also supported by the U.S. Department of Energy at Farmingdale State College through Contract No. DE-EE0003228,000. The work was performed at the facilities at the Advanced Energy Research & Technology Center (AERTC), SBU; Brookhaven National Laboratory; the Institute for Research and Technology Transfer at Farmingdale State College .

361361 Production of Hydrogen from High BTU Natural Gas By Non-Thermal Plasma

Abstract:

Ceramatec has been developing the use of non-thermal plasma to catalyze the reformation of natural gas and other hydrocarbons for several years. The use of the plasma to catalyze reformation of natural gas, JP-8, S-8, NATO F-76, bio-gas, and other hydrocarbons has been proven in the laboratory and in field operations. The Ceramatec plasma catalyzed reformer has been operated successfully in partial oxidation, auto-thermal, and steam reforming modes. The high voltage, low amperage plasma generated requires little power to operate. Recently, Ceramatec was contracted to provide five (5) plasma reformers for an ammonia plant in South America. Each reformer will produce ~ six (6) metric tons per day of hydrogen at about 50 psi. The units will operate in an auto-thermal mode with either air, or oxygen enriched air, being used as the oxidant. The proposed presentation will provide the advantages of the non-thermal plasma reformer, discuss past laboratory and field results, and describe the new work on the units being manufactured for ammonia production.

Kinetic modeling of decomposition reactions in novel microplasma reactors

Dr. Peter J. Lindner[1], Dr. Ronald S. Besser[2]

[1]Department of Chemical Engineering
Manhattan College
4513 Manhattan College Parkway
Riverdale, NY 10471

[2]Department of Chemical Engineering and Material Science
Stevens Institute of Technology
One Castle Point on Hudson
Hoboken, NJ 07030

Portable power generation technology has not been able to keep pace with the rising advances in the technology that requires it. Batteries have been the only suitable source for power in compact systems despite their low energy densities (~0.5 MJ/kg). The logical progression is to use high energy-dense fuels such as hydrogen (120 MJ/kg) to power efficient portable fuel cells. Fuel cells have advantages such as high energy densities, no moving parts, room temperature operation, and few undesirable products. The major drawback of fuel cells for portable power to date has been a readily available source of hydrogen. Hydrogen can be produced in large facilities but due to its low density it is logistically impractical to store and transport. Instead, it is preferable to use readily available hydrocarbon fuels to develop hydrogen on site. This can be done using common catalytic reactions, but catalysts generally require high temperatures, long processing time, and catalyst sites are often blocked with solid carbon and sulfur components which are commonly present in hydrocarbon reforming. As an alternative, a novel microplasma reactor was used to decompose hydrocarbons into their basic components.

These MEMS-based reactors have been successful in reforming fuels such as methane, acetylene, butane, methanol, and carbon dioxide but still suffer from sub-optimal efficiency. Analysis has shown that design improvements can increase the device lifetime and efficiency. In addition, the test system has inadequacies that tend to mask the intrinsic device performance. The system has a large excess volume limiting the microchannel reaction zone to only a small fraction of the total volume (1/8000). Despite these limitations, observed conversions have been relatively high (~50%).

Before the system is optimized for conversion and energy efficiency, it is necessary to better understand the reaction kinetics as a function of the input electrical power. In this study the kinetics of the microplasma reactions are modeled in COMSOL Multiphysics. A three-dimensional rendering of the void volume was developed matching the specific geometry of each reactor tested. Both steady state and unsteady state models are used with physics of laminar flow and transport of dilute species incorporated. By varying the reaction zone extent, reaction rates can be determined by matching the conversion with experimental data. These reaction rates are then compared with the input electrical power, and the specific energy input (SEI), to determine a correlation between electrical power and the reaction rate. After rates are determined as a function of electrical power or SEI, the inlet flow rates, excess volume, and reactant composition can be varied in the simulation to determine key system parameters for optimization.

This study provides a model approach for determining reaction kinetics for MEMS reactors and novel microplasma reactions in which physical details (such as microscale length and volume) of the reactor cannot be ignored and yet are difficult to assess directly. We will present the computational fluid dynamics model, the results of various reforming experiments, findings from iterations of the model (reaction rates, orders, etc.), and an improved system design that results. Future schematics of this design will also be presented with the aim of developing an integrated fuel processor-fuel cell system.

STEAM REFORMING OF BIO-OIL MODEL COMPOUNDS FOR SUSTAINABLE HYDROGEN PRODUCTION

Kepa Bizkarra[1], Victoria L. Barrio[1], Jesús M. Requies[1], Pedro L. Arias[1], José F. Cambra[1]

School of Engineering, University of the Basque Country UPV/EHU, Bilbao, Spain

Introduction

Hydrogen production has attracted great interest as a future clean energy fuel because of its potential applications in transportation and production of electricity. Nevertheless, in order to avoid CO_2 emissions during hydrogen production processes [1], the use of sustainable resources (biomass or bio-fuels) for hydrogen production may represent a good option to partially satisfy the energy demand [2], as it is considered renewable, available, and carbon-neutral source.

In this sense, biomass derived bio-oils are easily obtained and are recognized as appropriate fuels for hydrogen production [1]. Bio-oils are a complex mixture of high and low molecular weight, single and multifunctional hydrocarbons [3]. Therefore, in order to gain further insights into bio-oil reforming processes, model compounds have been widely studied in steam reforming conditions, like acetic acid [4], ethanol [2], and butanol [3], among others, as well as their thermodynamic analysis [5].

Nickel based catalysts are widely used in steam reforming because metallic nickel activates the organic molecule, breaking O-H, C-C and C-H bonds and promotes the reaction of organic fractions with the OH groups of water [6], in addition they show high selectivity to hydrogen and their cost is low when compared to noble metal based catalysts [7,8].

In this work butanol and m-xylene were selected as bio-oil representative model compounds (oxygenated and non-oxygenated) to be studied during the Steam Reforming (SR) process. For this purpose, Ni based systems using doped alumina were used. But the main goal of this work is the study of the best catalytic systems for the butanol and m-xylene steam reforming process in order to produce hydrogen. Catalysts were characterized using inductively coupled plasma atomic emission spectrometry (ICP-AES), N_2 physisorption, temperature programmed reduction (TPR), H_2 chemisorption, X-ray diffraction (XRD) and X-ray photoelectron spectroscopy (XPS) techniques.

Experimental procedure

Catalysts preparation and characterization

Wet impregnation was used to prepare nickel catalysts and modify supports. Catalysts were prepared to reach a 13 wt% of nickel and 10, 6 and 3 wt% of CeO_2, La_2O_3 and MgO, respectively. Catalysts with these composition showed a high hydrogen yield and stability during the SR of ethanol [9].

Commercial γ-alumina was sieved to a particle diameter between 0.42 and 0.50 mm and calcined at 973 K during 4 hours prior to the modification. Cerium (III) nitrate hexahydrate (99.5 %; Alfa Aesar), Lanthanum (III) nitrate hexahydrate (99.9 %; Alfa Aesar) and Magnesium nitrate hexahydrate (99.0 %; Fluka) were used as promoters. A solution of 10 mL of deionized water per gram of alumina was prepared and the measured amount of promoter was added for the support preparation. The resultant mixture was mixed and stirred overnight by means of a rotator evaporator. Afterwards, the solution was heated up to 338 K and water was removed with the aid of a vacuum pump. Then, supports were dried at 373 K and calcined at 973 K during 1 and 4 hours, respectively.

Nickel impregnation was carried out using Nickel (II) nitrate hexahydrate (99.99 %; Sigma Aldrich) as promoter. An aqueous solution of support was prepared taking 10 mL of deionized water per gram of support (bare alumina or modified alumina) and nickel precursor was added. The final solution was mixed, stirred, dried and calcinated as described in the previous paragraph.

Fresh catalysts textural properties were evaluated by means of N_2 adsorption-desorption isotherms obtained at 77 K using an Autosorb 1C-TCD. An inductively coupled plasma atomic emission spectroscopy (ICP-AES) instrument, model 2000-DV (Perkin Elmer), was used to determine the metallic elements in the fresh catalysts.

Temperature programmed reduction (TPR) was carried out using an Autosorb 1C-TCD apparatus, equipped with a thermal conductivity detector, and H_2-pulse chemisorption at 313 K in the Autosorb 1C-TCD unit was used to measure nickel dispersion, assuming an adsorption stoichiometry of H/Ni = 2 was assumed.

X-ray photoelectron spectroscopy, XPS, patterns of reduced and used catalysts were obtained using a SPECS (Berlin, Germany) system equipped with a Phoibos 150 1D-DLD analyzer and a Al Kα (1486.6 eV) monochromatic radiation source. Finally, X-ray powder diffraction, XRD, patterns were also collected using a PHILIPS X'PERT PRO automatic diffractometer operating at 40 kV and 40 mA, in theta-theta configuration, secondary monochromator with Cu-Kα radiation (λ = 1.5418 Å) and a PIXcel solid state detector.

Activity measurements

Activity tests were carried out using a Microactivity Reference (PID Eng&Tech) bench scale plant and two gas cromatographs for gas and liquid product analysis. A 316 stainless-steel fixed bed reactor (9 mm i.d. and 30.5 cm length, Autoclave Engineers) was used in all the experiments, as a conventional reaction system.

400 mg of catalyst with a particle size between 0.42 and 0.50 mm were used in each test. Catalyst was mixed with inert silicon carbide, SiC, (0.50-1.0 mm particle size) in a 1 to 9 w/w proportion, in order to avoid temperature gradients in the catalytic bed [10].

The fed gaseous mixture flows were adjusted by electronic controllers and two HPLC-Gilson liquid pumps were used for the bio-oil model compound and deionised water injection. The reaction products were cooled in a partial condenser. Condensed liquid was collected, weighted and offline analyzed with a gas chromatograph equipped with TDC and FID detectors. The gas phase was online analyzed with another gas chromatograph equipped with the same detectors. The use of this equipment allowed measuring different parameters, defined as:

n-butanol conversion:	X_{C4H10O} (%) = $(V_{C4H10O}^{in} - V_{C4H10O}^{out}) / V_{C4H10O}^{in} \cdot 100$
m-xylene conversion	X_{C8H10} (%) = $(V_{C8H10}^{in} - V_{C8H10}^{out}) / V_{C8H10}^{in} \cdot 100$
Hydrogen yield:	H_2 yield (%) = $V_{H2}^{out} / V_{H2}^{out}$ stoichiometric $\cdot$ 100
CO yield:	Y_{CO} (%) = $(V_{CO}^{out}) / V_{carbon\ fed} \cdot 100$
CO_2 yield:	Y_{CO2} (%) = $(V_{CO2}^{out}) / V_{carbon\ fed} \cdot 100$
CH_4 yield:	Y_{CH4} (%) = $(V_{CH4}^{out}) / V_{carbon\ fed}$ 100
C_2, C_3, C_4 yield:	$Y_{C2, C3, C4}$ (%) = $(V(_{C2+C3+C4})^{out}) / V_{carbon\ fed} \cdot 100$

Where: V_i^{in} corresponds to the mole flow-rate of reactant *i* (mol/min).

V_i^{out} corresponds to the mole flow-rate of product *i* (mol/min).

Before running the activity tests, each catalyst was reduced at 800 ºC, using 225 NmL/min of a 4:1 N_2:H_2 mixture during 4 h. All tests were carried out at atmospheric pressure and at a Steam to

Carbon (S/C) ratio of 5.0 and different reaction temperatures, from 1073 K to 873 K. All the tests lasting 4 hours to check which conditions are optimum to obtain the highest hydrogen yields and to make comparisons between catalysts.

Results and discussion

Catalysts textural properties and chemical composition

First, all catalysts were characterized using ICP-AES technique to determine their metal content, shown Table 1.

Table 1. Catalysts textural properties and chemical compositions.

	Surface area (m^2/g)	Pore volume (cm^3/g)	Pore size (Å)	Chemical composition Nominal / Measured by ICP-AES (wt. %)			
				Ni	CeO_2	La_2O_3	MgO
Alumina	202	0.81	77				
CeO_2-Al_2O_3	145	0.61	80				
La_2O_3-Al_2O_3	179	0.71	76				
MgO-Al_2O_3	179	0.72	77				
Commercial	22	0.09	170	13.0/12.4	-	-	-
Ni/Al_2O_3	147	0.55	72	13.0/13.9	-	-	-
Ni/CeO_2-Al_2O_3	127	0.49	78	13.0/13.7	10.0/10.9	-	-
Ni/La_2O_3-Al_2O_3	139	0.56	78	13.0/12.9	-	6.0/5.7	-
Ni/MgO-Al_2O_3	145	0.59	78	13.0/13.1	-	-	3.0/2.9

Supports and catalysts textural properties showed a reduction of the surface area while different compounds are impregnated in alumina. In a similar way, pore volume was also reduced with the impregnations. However, the average pore size remains constant, approximately. These results could be due to the pore blocking during the impregnations.

According with ICP-AES measurements, catalyst metal content is very close to the desired composition, for all metal or metal oxides present.

Reduced catalysts characterization

Figure 1 contains the TPR patterns of different catalysts and supports. Commercial catalyst showed four different reduction peaks around 775, 900, 1050 and 1150 K, while the rest of the catalysts only showed a wide reduction peak from 800 to 1300 K, approximately. However, in these broad peaks the reduction peaks of NiO, complex NiO_x with strong interactions with support material, and the reduction of $NiAl_2O_4$ [11] are present.

In the samples without nickel only CeO_2-Al_2O_3 showed reduction peaks around 800 and 1175 K, related with surface and bulk ceria reduction respectively [12,13].

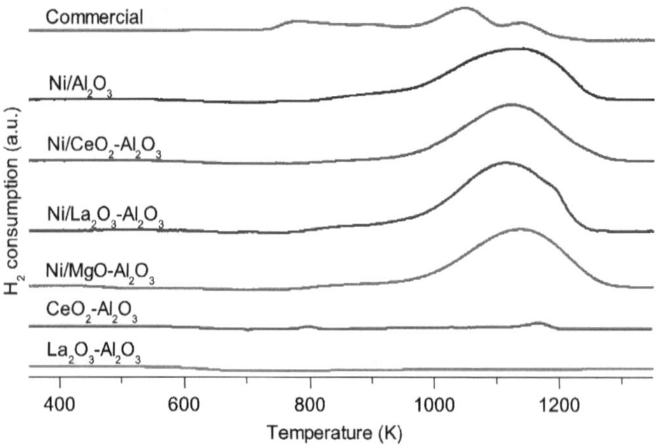

Figure 1. TPR profiles of the calcined fresh catalysts.

Table 2 contains the results of the hydrogen pulse chemisorptions. The amount of exposed nickel in all prepared catalysts was similar regardless of the support modifiers. Low nickel dispersion was observed in all catalysts, which is related with the low active metal surface area and high crystal sizes.

Table 2. Reduced catalysts H2 chemisorption results.

	Active metal surface area (m^2/g)	Metal dispersion (%)	Average Ni crystal size (nm)
Ni/Al_2O_3	4.3	4.7	22
$Ni/CeO_2-Al_2O_3$	4.4	4.9	21
$Ni/La_2O_3-Al_2O_3$	3.9	4.5	22
$Ni/MgO-Al_2O_3$	3.8	4.4	23

XRD technique was used to determine different crystallographic species in catalysts and to determine their average size through the Scherrer equation. Nickel crystals (Powder Diffraction File, PDF: 00-001-1260) with an average size of 5 nm in all prepared catalysts and 20 nm in the case of the commercial catalyst were measured.

Regarding support modifier oxides, only CeO_2 crystals (PDF: 01-075-0151) were measurable with an average size of 10 nm. However, the presence of lanthana and magnesia was confirmed by the ICP-AES and XPS measurements.

Two different alumina crystals were identified with this technique. Prepared catalysts contained alumina crystals with an average size of 5 nm, while commercial catalyst showed corundum (PDF: 01-074-1124) and Al_2O_3 crystals (PDF: 00-009-0440) with an average size of 40 nm.

XPS technique was used to measure metal atomic ratios and determine the reduced species in the catalytic surface. Table 3 compares the atomic ratios measured by XPS and ICP-AES.

Table 3. Comparison between the surface atomic ratio, measured by XPS, and bulk atomic ratio, measured by ICP-AES.

	Atomic ratios Measured by XPS / Measured by ICP-AES			
	Ni/Al	Ce/Al	La/Al	Mg/Al
Commercial	0.16 / 0.13	-	-	-
Ni/Al_2O_3	0.11 / 0.15	-	-	-
Ni/CeO_2-Al_2O_3	0.19 / 0.17	0.07 / 0.05	-	-
Ni/La_2O_3-Al_2O_3	0.19 / 0.14	-	0.09 / 0.06	-
Ni/MgO-Al_2O_3	0.13 / 0.14	-	-	0.16 / 0.05

Nickel (II), as nickel aluminate, was measured as the main nickel specie on the catalysts surface with a binding energy of 855.4 eV, with their characteristic nickel satellite peaks around 861 eV. However, at 852.75 eV metallic nickel was also measured, reaching an amount around 12% in all catalysts. Oxidation states determination was impossible in Ni/La_2O_3-Al_2O_3 catalyst because of the interference between nickel and lanthana signals. Nevertheless, a peak corresponding with Ni 2p ½ electron was measured at 879.5 eV, which was associated to nickel aluminate satellite peak. Therefore, the peak at 871.6 eV could be the sum of the peaks associated to nickel aluminate and metallic nickel, because their binding energies correspond with that peak.

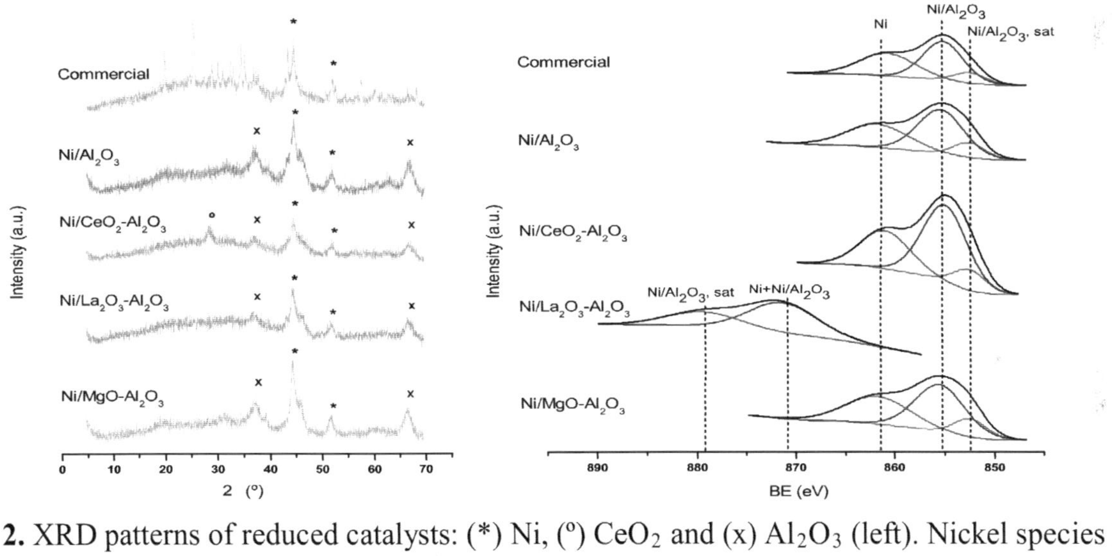

Figure 2. XRD patterns of reduced catalysts: (*) Ni, (°) CeO_2 and (x) Al_2O_3 (left). Nickel species XPS peaks decomposition for reduced catalysts (right).

Activity results

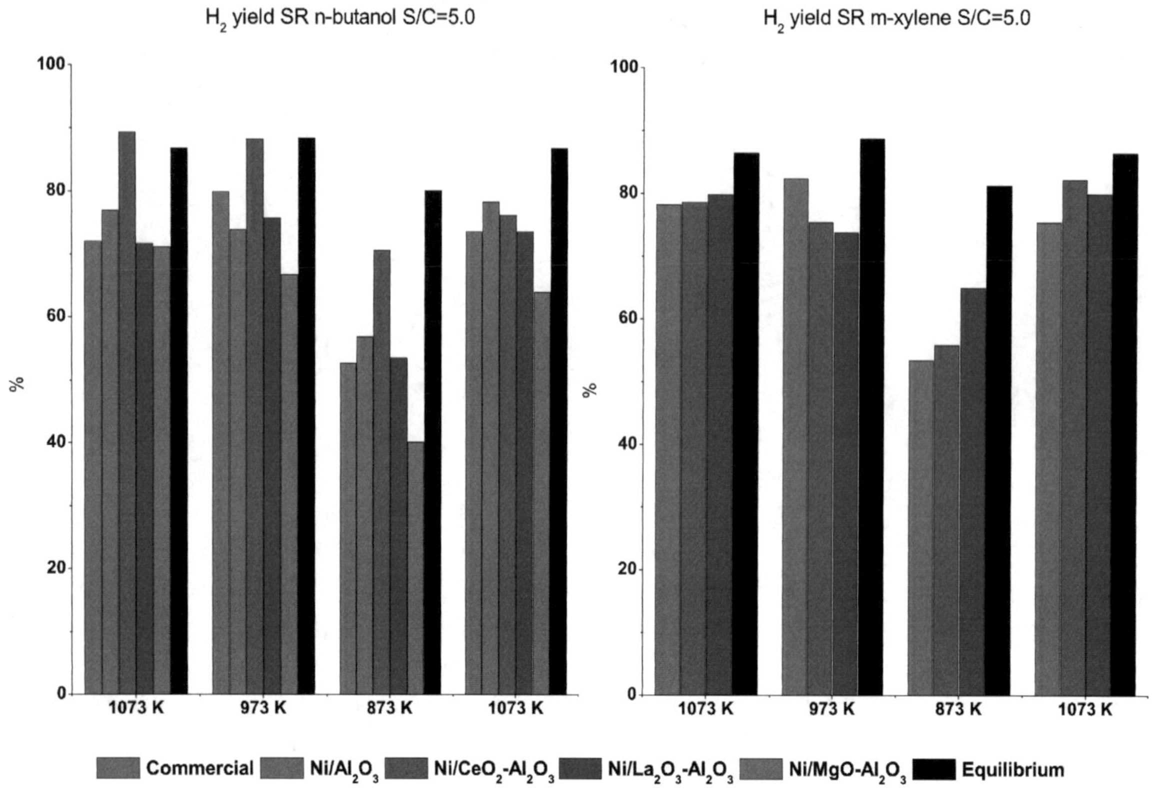

Figure 3. Comparison of hydrogen yields obtained with different catalysts during activity with n-butanol (left) and m-xylene (right).

Complete conversions were reached in all tested conditions with the two different bio-oil model compounds. Commercial catalyst was an exception because it reached a conversion higher than 95% during the SR of n-butanol at 873 K.

During the SR of n-butanol tests, the obtained hydrogen yield values were clearly lower than the equilibrium, in general terms, due to the presence of C_2, C_3 and C_4 species in the product gas, which do not appear in thermodynamic equilibrium conditions. Ni/CeO_2-Al_2O_3 catalyst showed the highest hydrogen yield in n-butanol SR at tested conditions.

Ni/Al_2O_3 catalyst achieved higher hydrogen yields than the commercial catalyst, except at 973 K. Support modification showed important differences in the catalytic performance. Ni/CeO_2-Al_2O_3 catalyst was able to achieve equilibrium hydrogen yield values at 1073 and 973 K, and at 873 K also was the catalyst which reached the highest hydrogen yield. However it showed deactivation evidences when the temperature was returned to the initial conditions. Despite the deactivation, ceria containing catalysts obtained one of the highest hydrogen yields at those conditions.

Ni/La_2O_3-Al_2O_3 catalyst showed a behavior similar to the commercial catalyst from the point of view of the hydrogen yield.

Ni/MgO-Al_2O_3 catalyst achieved the lowest hydrogen yields in all tested conditions. Besides it also showed deactivation evidences during the last reaction step at 1073 K.

In view of those results, commercial catalyst and Ni/MgO-Al_2O_3 catalysts were discarded for the next reaction series.

Tests carried out in the steam reforming of m-xylene achieved hydrogen yields close to the equilibrium values, due to the low amount of hydrocarbons in the product gas. It can be seen that the biggest difference between the equilibrium and experimental H_2 yields was obtained at 873 K, where around the 15% of the carbon fed was converted into hydrocarbons.

Alumina supported nickel based catalysts showed a similar behavior during the tests at 1073 K, regardless of the support modifications. They reached a hydrogen yield of 80%.

When reaction temperature was reduced to 973 K, bare alumina supported catalyst and the catalysts supported on CeO_2-Al_2O_3 maintained the hydrogen yield reached at 1073 K. However, Ni/La_2O_3-Al_2O_3 catalyst slightly reduced its hydrogen yield.

Nevertheless, lantana containing catalyst was the catalyst which showed the highest hydrogen yield during the tests at 873 °K, followed by the Ni/CeO_2-Al_2O_3 catalyst. In that conditions, bare alumina supported catalyst was which obtained the lower hydrogen yield value.

Comparing initial and final hydrogen yields at 1073 K, it can be observed how the catalysts with the support modified with a metal oxide maintain their initial activity after the tests at different temperatures. On the other hand, during the final reaction step at 1073 K, Ni/Al_2O_3 catalyst shows a hydrogen yield lower than the initial, which could be related to deactivation processes.

Conclusions

The main goal of this work is the valorization of the bio-oil for hydrogen production, in order to promote the intensive application of renewable resources like the biomass. Therefore two bio-oil representative model compounds were tested at S/C=5.0, atmospheric pressure and different temperatures.

In general, hydrogen yields obtained from m-xylene SR process are higher than the obtained from n-butanol SR, mainly because of the different amount of hydrocarbons in the product gas. Therefore it could be due to the higher reactivity of the aromatic compound in tested conditions.

On the other hand, the low hydrocarbon content measured during m-xylene SR test could improve the stability of the catalysts because ethylene is an important carbon precursor. That explains the low deactivation evidences of the tested catalysts.

It is remarkable the different performance of the catalyst depending on the model compound fed. When butanol is fed to the system, at 600 °C the CeO_2 doped catalyst is the most active, while the La_2O_3 modified catalyst is the most active in the m-xylene reaction.

When higher temperature is employed, the activity and hydrogen yield is almost the same in both cases.

Extensive catalyst characterization will be done, and it will be correlated with the activity results.

References

1. Hu, Xun, Lu, Gongxuan. (2009) "Inhibition of methane formation in steam reforming reactions through modification of Ni catalyst and the reactants", Green Chemistry. pp. 724-732.
2. Benito, M., Padilla, R., Serrano-Lotina, A., Rodríguez, L., Brey, J. J. and Daza, L. (2009) "The role of surface reactions on the active and selective catalyst design for bioethanol steam reforming", Journal of Power Sources. pp. 158-164.
3. Bimbela, F., Oliva, M., Ruiz, J., García, L. and Arauzo, J. (2009) "Catalytic steam reforming of model compounds of biomass pyrolysis liquids in fixed bed: Acetol and n-butanol", Journal of Analytical and Applied Pyrolysis. pp. 204-213.

4. An, Lu, Dong, Changqing, Yang, Yongping, Zhang, Junjiao and He, Lei. (2011) "The influence of Ni loading on coke formation in steam reforming of acetic acid", Renewable Energy. pp. 930-935.
5. Aktaş, Seda, Karakaya, Mustafa and Avcı, Ahmet K. (2009) "Thermodynamic analysis of steam assisted conversions of bio-oil components to synthesis gas", International Journal of Hydrogen Energy. pp. 1752-1759.
6. Sánchez-Sánchez, M. C., Navarro, R. M. and Fierro, J. L. G. (2007) "Ethanol steam reforming over Ni/La–Al2O3 catalysts: Influence of lanthanum loading", Catalysis Today. pp. 336-345.
7. Medrano, J. A., Oliva, M., Ruiz, J., García, L. and Arauzo, J. (2009) "Catalytic steam reforming of model compounds of biomass pyrolysis liquids in fluidized bed reactor with modified Ni/Al catalysts", Journal of Analytical and Applied Pyrolysis. pp. 214-225.
8. Iriondo, A., Barrio, V. L., Cambra, J. F., Arias, P. L., Güemez, M. B., Navarro, R. M., Sanchez-Sanchez, M. C. and Fierro, J. L. G. (2009) "Influence of La2O3 modified support and Ni and Pt active phases on glycerol steam reforming to produce hydrogen", Catalysis Communications. pp. 1275-1278.
9. Sánchez-Sánchez, M. C., Navarro, R. M. and Fierro, J. L. G. (2007) "Ethanol steam reforming over Ni/MxOy-Al2O3 (M=Ce, La, Zr and Mg) catalysts: Influence of support on the hydrogen production", International Journal of Hydrogen Energy. pp. 1462-1471.
10. Izquierdo, U., Barrio, V. L., Lago, N., Requies, J., Cambra, J. F., Güemez, M. B. and Arias, P. L. (2012) "Biogas steam and oxidative reforming processes for synthesis gas and hydrogen production in conventional and microreactor reaction systems", International Journal of Hydrogen Energy. pp. 13829-13842.
11. Koo, Kee Y., Roh, Hyun-Seog, Seo, Yu T., Seo, Dong J., Yoon, Wang L. and Park, Seung B. (2008) "Coke study on MgO-promoted Ni/Al2O3 catalyst in combined H2O and CO2 reforming of methane for gas to liquid (GTL) process", Applied Catalysis A: General. pp. 183-190.
12. de Freitas Silva, T., Dias, J. A. C., Maciel, C. G. and Assaf, J. M. (2012) "Ni/Al2O3 catalysts: effects of the promoters Ce, La and Zr on the methane steam and oxidative reforming reactions", Catalysis Science & Technology. pp. 635-643.
13. Vagia, Ekaterini C., Lemonidou, Angeliki A. (2010) "Investigations on the properties of ceria–zirconia-supported Ni and Rh catalysts and their performance in acetic acid steam reforming", Journal of Catalysis. pp. 388-396.

391904 Micro-Structured Reactors for Fuel Processing

Abstract:

In parallel to the establishment of micro-reactors in the field of liquid phase chemistry, gas phase reactions have been the subject of numerous investigations. This has quickly led to the introduction of heterogeneous catalysts as coatings or fixed beds into the reactors.

The process intensification potential of micro-structured reactors has attracted attention in the field of energy technology, wherever compact, decentralized solutions were required. Meanwhile, the entire field from portable and mobile power generation to stationary and decentralized production of fuels and power is subject of worldwide research efforts in the field of micro-technology. One prominent example are auxiliary power units (APU) based upon fuel cell technology for applications such as aircrafts, trucks, recreational and other vehicles. To meet the stringent size and cost demands of such applications the step towards micro-structured plate heat-exchanger technology was an obvious one. The number of customers addressed by mobile power generation systems create a mass market, ranging from hundreds of units per year to tens of thousands.

The presentation will provide an overview of developments in the field focussing on recent work and practical applications.

384886 Annular Micro-Channel Reactors; Efficent and Scalable CH4-Steam Reformers

Abstract:

The experimental data, current development along with simulations as well as future design concepts will be discussed for the annular micro-channel reactor (AMR). This presentation will encompass design, construction and experimental testing that has been done to date. We will explain how this system could improve the hydrogen production industry using current resources and infrastructure. Experimental data from different catalyst thicknesses as well as flow rates will be shown. The AMR is a novel design for an efficient and compact methane-steam reformer to produce hydrogen that is relatively inexpensive to build. It consists of a tube in tube design with catalyst affixed to the ID of the outer tube. The AMR has the potential to produce hydrogen more directly at fueling stations due to the compact size and cost. The AMR would lessen the need for hydrogen storage as the reactor can be turned up and down to meet current demand. The AMR features a tube in tube design where a micro-channel exists between the two tubes. This is where a catalytic reaction occurs with the steam and methane to yield hydrogen. The AMR reforms hydrogen more efficiently than current steam reformers. We will show how this technology is also scalable allowing the AMR to be tailored to the amount of demand it might see. This flexibility is what gives the AMR great potential to be used in many more applications than current steam-methane reformers.

Sorption-Enhanced Variable Volume Membrane Reactor for Hydrogen Production from Methane: Modeling and Experimental Characterization

David M. Anderson[1], Peter A. Kottke[1], Thomas Yun[1] and Andrei G. Fedorov[1,2,*]

1. G.W. Woodruff School of Mechanical Engineering, Georgia Institute of Technology, Atlanta Georgia, 30332

2. Parker H. Petit Institute of Bioengineering and Bioscience, Georgia Institute of Technology, Atlanta Georgia, 30332

*Correspondence: AGF@gatech.edu

Abstract

Abundant petroleum and coal have enabled the production of affordable primary power and transportation fuels over the past century; however, the recent resurgence of natural gas as a significant fossil fuel resource has dramatically changed the energy landscape.[1] With its favorable hydrogen-to-carbon ratio coupled with recently developed techniques to tap vast reserves that were previously not economically viable, natural gas has been touted by many as a bridge fuel to a low-carbon energy future.[2] To make full use of this promising resource, new chemical conversion processes which are optimized for natural gas feedstock must be developed. A process that can efficiently produce hydrogen in a small-scale, distributed fashion from methane, the primary component of natural gas, is one such technology that would have broad-reaching impacts, as it could eliminate the requirement for a large-scale hydrogen delivery infrastructure and aid in the diffusion of hydrogen-powered technology.

Hydrogen production via the steam-methane reforming (SMR) process is a mature industrial technology;[3] yet miniaturizing this process for distributed production has proven problematic. Owing to its strongly endothermic nature, reaction temperatures in excess of 900°C are traditionally employed in large-scale SMR reactors to overcome the thermodynamic limitations of the process.[4] For a distributed reactor system, an operating temperature below 500°C is desired to improve safety and reduce material cost and preheating energy requirements. Furthermore, multiple sequential unit operations are required after the primary reforming reaction step to convert unwanted CO into H_2 through the water gas shift (WGS) reaction and purify the H_2 product via adsorption separation techniques.[4] _ENREF_8 This scaled-down sequential unit operation approach requires an unsatisfactorily large footprint and results in system losses and balance-of-plant component costs becoming a larger fraction of the overall output.

Combined reaction-separation approaches are a promising means to overcome the SMR process scale-down limitations. By Le Chatelier's principle, *in situ* removal of H_2 and/or CO_2 from the reactor shifts equilibrium toward greater fuel conversion and H_2 yield, producing a suitable level of conversion at much lower temperature and enabling removal of downstream WGS and purification units.[5] To that end, this work investigates a CO2/H2 Active Membrane Piston (CHAMP) variable volume, batch-membrane reactor which incorporates selective CO_2 adsorption and enable distributed H_2 production. Prior work on his concept, termed CHAMP-SORB, has established the thermodynamic viability of this approach and motivated a more detailed theoretical study of reaction kinetics and experimental validation.[6]

The basic CHAMP reactor, operating in a cycle similar to an Internal Combustion (IC) engine, has proven capable of dynamically maintaining optimal reaction conditions for transport of reactants to the catalyst, reaction kinetics at the catalyst, and H_2 membrane separation.[7] The CHAMP-SORB advances this concept further by employing a mixture of SMR catalyst and solid CO_2 sorbent to provide additional enhancement of the forward SMR reaction. Figure 1~~Figure 1~~ illustrates the basic design and operation of the CHAMP-SORB reactor. In step 1, the reactor inlet valve is opened and the piston is moved upward to intake a mixture of CH_4 and H_2O fuel. Once the reactor is filled, step 2 consists of transport of the fuel to the catalyst/sorbent admixture for conversion into products (H_2, CO_2 and CO). As this conversion occurs, H_2 is removed from the reactor chamber via permeation through an infinitely selective palladium-silver membrane and CO_2 is removed from the gas phase via adsorption. Low partial pressure of H_2 on the permeate side of the membrane is maintained via an Argon sweep gas or through consumption in an electrochemical device. As fuel is depleted and the reactor pressure is reduced due to permeation/adsorption, the driving force for reaction, permeation and adsorption is restored through downward motion of the piston. In step 3, the outlet valve is opened and the remaining non-adsorbed CO_2, as well as unreacted fuel and trace CO/H_2, is exhausted from the reactor. To prevent saturation of the sorbent for subsequent cycles, regeneration is accomplished in step 4 by closing the inlet/outlet valves and elevating the piston to expand the reactor volume, thereby creating favorable conditions for CO_2 desorption. Lastly, the outlet valve is reopened and the piston moves downward to exhaust the desorbed CO_2 and prepare the reactor for its subsequent cycle.

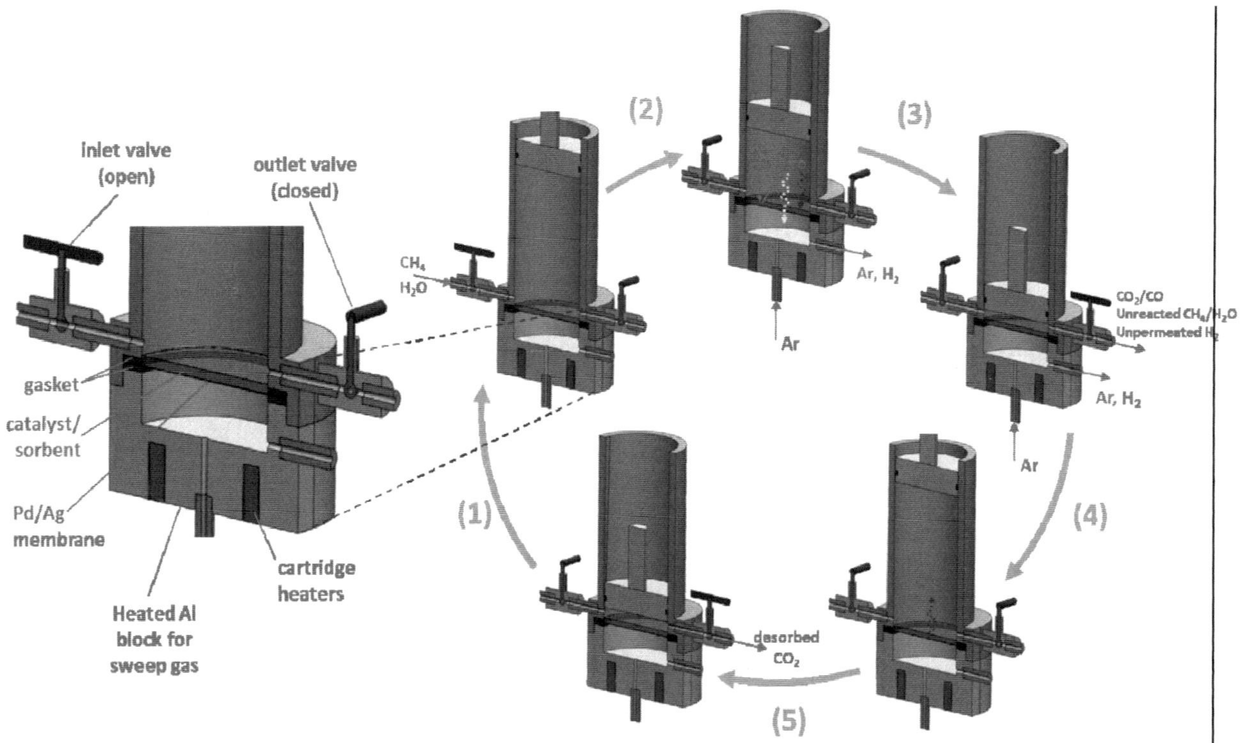

Figure 1: CHAMP-SORB design and operational schematic

In this work, a comprehensive reactor model is developed to explore the CHAMP-SORB reactor concept. Reaction kinetics from Xu/Froment[8] and Hou/Hughes[9] are employed in the model, as well as adsorption isotherm data and linear driving force kinetics for potassium-promoted hydrotalcite from Lee et al.[10] The model accounts for species transport via advection and multicomponent diffusion to/from the catalyst/sorbent admixture. Heat transfer to the catalyst/sorbent is also modeled as convection from the preheated Argon sweep gas flowing through the heated aluminum bottom block. Membrane permeation is modeled as proportional to the H_2 partial pressure difference across the membrane according to Sievert's law, accounting for a reduction in permeation due to competitive adsorption of other species.[11] To model the transient profile of H_2 partial pressure on the membrane permeate side, the Ar/H_2 mixture is approximated as a well-mixed at the outlet. Specifically investigated are the relative impacts of catalyst-to-sorbent and steam-to-carbon ratios, as well as the operating pressure and temperature in both the reaction and regeneration steps. It is found that an ideal catalyst-to-sorbent ratio of approximately 1:4, depending on other specific operating conditions, produces an optimal balance of average H_2 yield rate and fuel utilization. Additionally, it is possible to achieve fuel utilization in excess of 95% at the stoichiometric steam-to-carbon ratio of 2:1, which aids in maximizing H_2 yield density and thermal efficiency by avoiding the energetic penalty associated with creation of and reactor volume used to hold excess steam.

Lastly, experimental results with a fully operational CHAMP-SORB testbed are reported. The H_2 content from the sweep gas as well as the quantities of unreacted, non-permeated and non-adsorbed species are measured by a Hiden Analytical Quadrapole HPR-20 mass spectrometer. The reactor pressure is measured by a high temperature pressure transducer (Kulite Semiconductor Products, XTEH series) and utilized in a PID control loop actuate the piston via a Newmark Systems linear actuator. The experimental results are used to compare with and better understand the comprehensive reactor model.

References

1. Rao V. Shale Gas, The Promise and the Peril: RTI Press; 2012.

2. Paltsev S, Jacoby HD, Reilly JM, et al. The future of US natural gas production, use, and trade. *Energy Policy.* 2011;39(9):5309-5321.

3. Rostrup-Nielsen J. Steam reforming of hydrocarbons. A historical perspective. *Natural Gas Conversion Vii.* 2004;147:121-126.

4. Harrison DP. Sorption-enhanced hydrogen production: A review. *Ind Eng Chem Res.* Sep 2008;47(17):6486-6501.

5. Lee KB, Beaver MG, Caram HS, Sircar S. Novel thermal-swing sorption-enhanced reaction process concept for hydrogen production by low-temperature steam–methane reforming. *Ind Eng Chem Res.* 2007/07/01 2007;46(14):5003-5014.

6. Anderson DM, Kottke PA, Fedorov AG. Thermodynamic analysis of hydrogen production via sorption-enhanced steam methane reforming in a new class of variable volume batch-membrane reactor. *Int J Hydrogen Energy.* 2014(In press).

7. Damm DL, Fedorov AG. Batch reactors for hydrogen production: Theoretical analysis and experimental characterization. *Ind Eng Chem Res.* Jun 17 2009;48(12):5610-5623.

8. Xu JG, Froment GF. Methane steam reforming, methanation and water-gas shift 1. Intrinsic kinetics. *AIChE J.* Jan 1989;35(1):88-96.

9. Hou K, Hughes R. The kinetics of methane steam reforming over a Ni/α-Al< sub> 2</sub> O catalyst. *Chemical Engineering Journal.* 2001;82(1):311-328.

10. Lee KB, Verdooren A, Caram HS, Sircar S. Chemisorption of carbon dioxide on potassium-carbonate-promoted hydrotalcite. *J Colloid Interf Sci.* Apr 2007;308(1):30-39.

11. Israni SH, Harold MP. Methanol steam reforming in Pd– Ag membrane reactors: effects of reaction system species on transmembrane hydrogen flux. *Ind Eng Chem Res.* 2010;49(21):10242-10250.

392014 A Novel TPOX-Based "Swiss-Roll" Fuel Reformer

Abstract:

A compact, non-catalytic, thermal partial oxidation (TPOX) based reformer has been developed to provide hydrogen-rich syngas to solid oxide fuel cells (SOFC) using fuel-rich hydrocarbon-air mixtures as a feedstock. The reformer is essentially a spiral heat exchanger with chemical reaction and heat release occurring at the center. It is highly effective at recuperating heat from the hot reformate stream and transferring it to the inlet fuel-air stream such that superadiabatic temperatures can be achieved within the central heat release / reforming zone. These high temperatures combined with long residence times in the preheated inlet channel enable the reformate to approach the equilibrium composition (the maximum syngas yield) without the need for catalysts. The reaction is self-sustaining and does not require external energy other than for startup. Syngas production has been demonstrated using rich propane-air, n-heptane-air, and JP-8-air premixed feedstocks over a relatively broad range of operating conditions. Representative results will be presented for select fuels, inlet conditions, and reformer geometry.

392069 Characterization and Mechanistic Study of Dry Reforming of Methane over Rh Substituted Lanthanum Zirconate (La2Zr2O7) Pyrochlores

Abstract:

Dry reforming of methane (DRM) to syngas (H_2 and CO) has been well-studied over a number of catalysts. The high temperatures needed to produce syngas inevitably result in carbon formation. Although pyrochlores are known to be thermally stable at conditions required for dry reforming, there is very little literature on this reaction for these catalysts. Catalytically active metals like Rh and Ru can be isomorphically substituted into the crystalline structure of the pyrochlores to produce stable catalyst. These materials also resist carbon formation due to their inherent oxygen conductivity at conditions required for DRM. Here, we report the use of Rh-substituted- lanthanum zirconate ($La_2Zr_2O_7$) pyrochlores for DRM. Specifically three catalysts (a) unsubstituted lanthanum zirconate (LZ) for baseline comparison, (b) 2% and (c) 5% (by wt) Rh-substituted L2RhZ and L5RhZ pyrochlores were characterized using X-ray diffraction, CH_4 and H_2 temperature programmed reduction (TPR). The most crucial step in DRM mechanism is the activation of CH_4. This step is reported to be the slowest and the rate determining step in the DRM mechanism. In this work we study this CH_4 activation step by means of Arrhenius plot, CH_4/CD_4 kinetic isotope effect.

383934 Oxygen-Enhanced Water Gas Shift over Bimetallic Catalysts for More Efficient Hydrogen Production

Abstract:

Water gas shift (WGS) and deep CO removal are critical for hydrogen production and low-temperature fuel cell applications. Our recent work demonstrates the significant advantage of oxygen-enhanced water gas shift (OWGS) for more efficient H_2 production over bimetallic catalysts, wherein a small amount of O_2 is added to reformate gas to enhance CO shift. Pd-Cu and Pt-Cu bimetallic catalysts supported on CeO_2 show strong synergetic promoting effect in OWGS, which leads to much higher CO conversion and higher H_2 yield than WGS at low temperature around 250 ^oC. TPR showed strong interaction between Pd and Cu in Pd-Cu/CeO_2 by a single reduction peak in contrast to multiple peaks on monometallic Cu/CeO_2. EXAFS analysis revealed that such bimetallic Pd-Cu and Pt-Cu form alloy nanoparticles, where noble metal is mainly surrounded by Cu atoms. OSC measurements point to higher resistance of Pd-Cu to oxidation indicating that Pd keeps Cu in reduced state in air pulse condition. From kinetic study, Pd in Pd-Cu was found to promote CO shift, rather than CO oxidation, and Cu in Pd-Cu suppress H_2 activation (that is inherent to monometallic Pd), which minimizes both the inhibition effect of H_2 and the loss of H_2 by oxidation in OWGS. Transient response technique revealed that Cu in Pd-Cu enhances desorption of strongly chemisorbed CO_2 on catalyst surface in contrast to very slow CO_2 desorption from surface of monometallic Pd. The role of CeO_2 support will also be discussed.

Fabrication of Compact Catalytic Membrane Reactor Using Nanosize Catalysts and Atomic Layer Deposition

Shuai Tan [1], Shannon Mahurin [2], Stephen N. Paglieri [3] and Dongmei Li [1]
[1] University of Wyoming, Laramie, WY, 82071. dli1@uwyo.edu
[2] Oak Ridge National Laboratory, Oak Ridge, TN, 37831. mahurinsm@ornl.gov
[3] TDA Research, Inc., Wheat Ridge, CO, 80033. spaglieri@tda.com

A novel protocol is presented for the development of a compact catalytic membrane reactor (CCMR) by impregnating nanosize sulfur tolerant water gas shift (WGS) catalysts into a porous composite layer supported on porous stainless steel (PSS). A submicron palladium (Pd) membrane was deposited on the opposite side of the PSS via Atomic layer deposition (ALD). Unlike existing catalytic membrane reactors with either packed bed or micro-channel configurations, this CCMR protocol allows for catalyst distribution inside and outside of the porous layer, maximizing WGS reactant contact area, while maximizing hydrogen (H_2) permeability by significantly reducing Pd membrane thickness. To prove the feasibility of nanocatalyst incorporation into a porous support, nano-lanthanide oxysulfide (Eu_2O_2S) particles were synthesized and impregnated into the porous ZrO_2 layer of AccuSep® (Pall Inc.), which is supported on PSS and has nominal pore size of 100 nanometers. To shift reaction thermodynamics toward hydrogen production with maximized H_2 permeability, Pd was deposited onto the inner surface of the porous stainless steel support using $Pd(hfac)_2$ and formalin as precursors via ALD. To enhance adhesion between the Pd membrane and the PSS support, an alumina (Al_2O_3) seed layer was deposited onto PSS. Preliminary testing of the supported nanocatalysts in a CCMR has shown comparable or better performance and stability for the WGS reaction in both sulfur-free and sulfur-laden (hydrogen sulfide, H_2S) environment, compared to commercial WGS catalysts, with the catalyst being fully recoverable from 121 ppm H_2S. It is notable that the CMMR protocol described here is not limited to Eu_2O_2S for WGS or ALD deposited Pd for hydrogen separation. It can be applied to incorporate various types of nanocatalysts and other ALD-deposited thin films for various gaseous compound generation and separation within one compact device, dramatically intensifying many chemical processes.

Key words: palladium, catalytic membrane reactor, atomic layer deposition, in-situ hydrogen production

Rational Design of Nano-Structured Oxygen Carriers for Chemical Looping Conversion of Methane

Fanxing Li
Department of Chemical and Bimolecular Engineering
North Carolina State University
Raleigh, NC 27695
* Email: fli5@ncsu.edu

As the primary component of natural gas, methane is among the most abundant organic compounds on earth. The current approach for liquid fuel and chemical production from methane involves methane reforming in the presence of gaseous oxidants such as steam, oxygen, and/or carbon dioxide. Although reforming based methods, particularly steam methane-reforming, have been successfully utilized in commercial processes, the efficiencies of state-of-the-art reforming processes are limited due to the high steam to methane ratio required by the reforming catalysts and/or the needs for energy intensive air separation operations. The current study investigates oxygen carriers, a.k.a redox catalysts, for converting methane into syngas and/or carbon dioxide in a chemical looping scheme. A number of redox catalysts composed of a primary metal oxide and mixed ionic-electronic conductor (MIEC) support, assembled as sub-micron composite or nano-sized core-shell particles, are synthesized and characterized. The mechanisms of the corresponding redox reactions are also investigated. Experimental results indicate that the core-shell redox catalysts are significantly more active than conventional oxygen carriers. Strategies for rational design and optimization of redox catalysts for chemical looping combustion and reforming schemes are also discussed based on these findings.

Encapsulated iron oxide nanoparticles for chemical looping based hydrogen production

Davood Hosseini, Christoph Müller
Laboratory of Energy Science and Engineering,
Department of Mechanical and Process Engineering, ETH Zurich, Switzerland

Chemical looping (CL) based hydrogen production involves a two-step process in which in the first step a synthesis gas reduces a metal oxide (Me_xO_y) oxygen carrier (Eqs. 1,2). Subsequently, the reduced metal oxide is re-oxidized using steam to produce a pure stream of H_2 (Eq. 3).

$$H_2 + Me_xO_y \leftrightarrow H_2O + Me_xO_{y-1} \quad (1)$$

$$CO + Me_xO_y \leftrightarrow CO_2 + Me_xO_{y-1} \quad (2)$$

$$H_2O + Me_xO_{y-1} \leftrightarrow H_2 + Me_xO_y \quad (3)$$

An appropriate oxygen carrier for the CL-based H_2 production must possess the following characteristics[1]:

(i) Thermodynamics that favor the complete conversion of synthesis gas to CO_2 and H_2O.

(ii) Regenerable with steam with a high 'hydrogen to steam' ratio.

(iii) Fast and stable kinetics over multiple cycles.

In this regard iron oxide is arguably the most promising material due to favorable thermodynamics, wide availability and being environmentally benign[1,2]. However, the lack of sufficient redox stability of pure Fe_2O_3 at high temperatures (700-1000 °C) is a major concern. Bohn et al.[3] demonstrated that unsupported Fe_2O_3 deactivates after only a few redox cycle. A possible solution to stabilize the redox performance of iron oxide is by supporting it on a high Tammann temperature material, such as SiO_2, TiO_2, ZrO_2, or CeO_2[4]. Within this strategy fall core-shell nanomaterials which have shown promising results in stabilizing the activity of various catalysts.[5]

Here we report the development of iron oxide nanoparticles encapsulated by SiO_2, CeO_2 and ZrO_2 for the chemical looping-based production of hydrogen. The materials were characterized by high resolution transmission electron microscope (HR-TEM), powder X-ray diffraction and N_2 adsorption. The redox performance of the synthesized oxygen carriers were assessed in a thermo-gravimetric analyser. The shell thickness is optimized to ensure a good compromise between redox stability and mass transfer characteristics. The newly developed materials show an improved stability against sintering when compared to conventionally supported oxygen carriers.

---

[1] P. Gupta, L. Velazquez-Vargas, L. Fan, Energy Fuels (2007) 21, 2900-8.
[2] R. D. Solunke and G. T. Veser, Ind. Eng. Chem. Res. (2010), 49, 11037.
[3] C. D Bohn, J. P. Cleeton, C.R. Müller, S. Y. Chuang, S. A. Scott, J. S. Dennis, Energy Fuels, (2010), 24, 4025-4033.
[4] J. Adanez, A. Abad, F. Garcia-Labiano, P. Gayan, L. F. de Diego, (2012), 38, 215-282.
[5] R. G. Chaudhuri and S. Paria, Chem. Rev. (2012), 112, 2373–2433

Fe and Ni-based Oxygen Carriers for CO_2 activation via Chemical Looping

Amey More, Saurabh Bhavsar and Götz Veser

Department of Chemical Engineering, University of Pittsburgh, Pittsburgh, PA

Abstract:

'Chemical Looping Combustion' (CLC) is an emerging clean combustion technology which offers an efficient and elegant route for fossil fuel combustion with inherent CO_2 capture based on the cyclic oxidation and reduction of an "oxygen carrier" (typically a metal) with air and a fuel, respectively. We have recently shown that this process can also be used for CO_2 activation by replacing air with CO_2 as the oxidant for the process, thus enabling the reduction of CO_2 to CO (which has many uses in the chemical industry) as an alternate pathway to carbon sequestration. We had identified Fe as the most suitable metal for the process due to its good reactivity with CO_2 at realistic process operating conditions. However, Fe shows poor reactivity with methane as a typical fuel for this process. Ni and NiO, on the other hand, is well known as an efficient catalyst for methane combustion, but cannot be oxidized with CO_2. We hence hypothesized that a bimetallic Fe/Ni oxide would show combined good reactivity with both CH_4 and CO_2 in the reduction and oxidation half cycles, respectively.

Supported Ni, Fe, and Ni/Fe oxygen carriers with 40wt% metal loading were synthesized using a simple wet-impregnation procedure. Three different ratios of the Fe to Ni were studied for their performance. Thermo-gravimetric analysis (TGA) of the carriers combined with X-Ray Diffraction (XRD) gave insight into oxidation states, carrier conversions, and reaction rates were evaluated in fixed-bed reactor experiments. We found that bimetallic oxides can indeed show activity for both methane combustion and CO_2 reduction; albeit the individual rates of oxidation are lower than for monometallic Fe carriers, and rates of reduction are lower than for monometallic Ni carriers. For comparison, physical mixtures of monometallic Fe and Ni carriers with the same metal weight loadings were prepared to isolate the oxidation characteristics of the individual metal oxides during the reactive processes. Remarkably, a significantly higher rate of carrier conversion obtained with the physical mixture—not only in comparison to the bimetallic carriers, but also in comparison to either monometallic carrier—suggested that the steps involved were quite different from those in case of a mixed oxide. The remarkable behavior can be explained based on a synergistic reaction sequence between the two carrier materials, which will be discussed in detail in the presentation. Overall, our study thus identifies a very simple, yet highly promising carrier system for an efficient chemical looping process for CO_2 activation.

383462 Continuous Regime of Clc/CLOU Reactivity and Its Impact on the Dynamic and Steady State Performance of Cu_xO_y Oxygen Carriers

Abstract:

A detailed kinetic network that captures the continuous regime of Cu-based oxygen carrier reactivity in chemical-looping combustion (CLC) and/or chemical-looping with oxygen uncoupling (CLOU) processes is used to study transient and steady state process efficiencies. Theoretical models for CLC/CLOU processes and experimental facilities are utilized for this analysis. The transient kinetics of Cu-based oxygen carriers in the temperature range of 750 to 1000 °C was investigated in a fixed-bed reactor (Figure 1), supported by supplementary characterization techniques, such as TGA, SEM and in-situ hot-stage XRD (Figure 2). The continuous regime between CLC and CLOU, as studied experimentally and theoretically is shown in Figure 3. The effect of operating pressure on the reaction kinetics was also investigated for the purpose of predicting syngas CL processes integrated in IGCC power/hydrogen plants. The established kinetic network is then implemented in different process models to evaluate system performance. The transient behavior of CLC/CLOU is simulated by a hydrodynamic three-phase (bubble, emulsion and wake) model in a fluidized bed reactor, while Aspen Plus process flowsheet simulations are performed to study the steady state behavior. On the basis of the developed Cu-based kinetics and process models, a comprehensive evaluation of the real efficiency of Cu-based CL processes integrated in the power generation infrastructure is carried out.

Acknowledgement: This material is based upon work supported by the National Science Foundation under Grant No. 1054718.

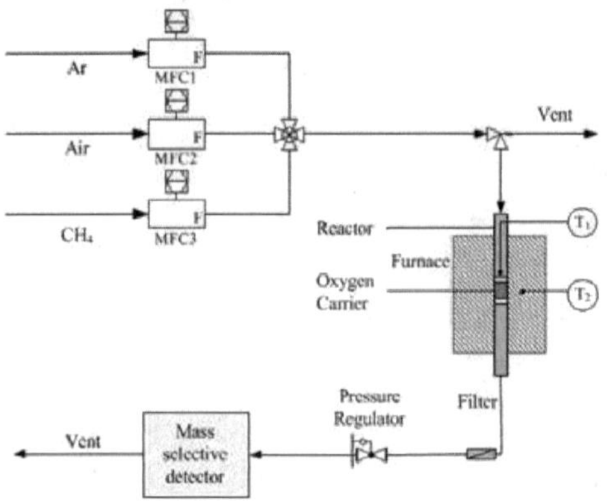

Figure 1: Fixed-bed chemical-looping setup used in this work.

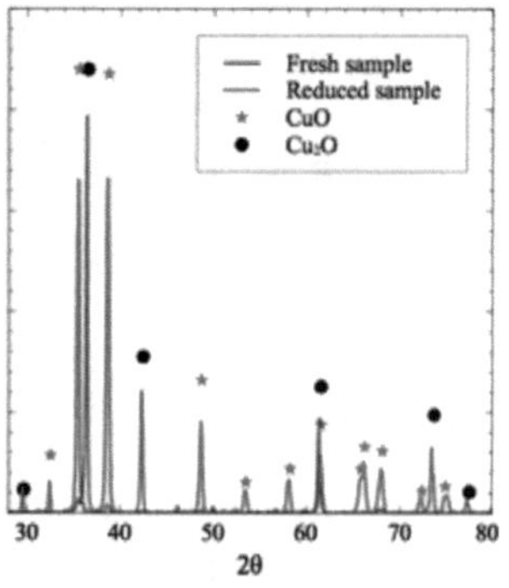

Figure 2: XRD spectra of SiO_2-supported CuO sample in oxidized and reduced (decomposed in Ar) forms.

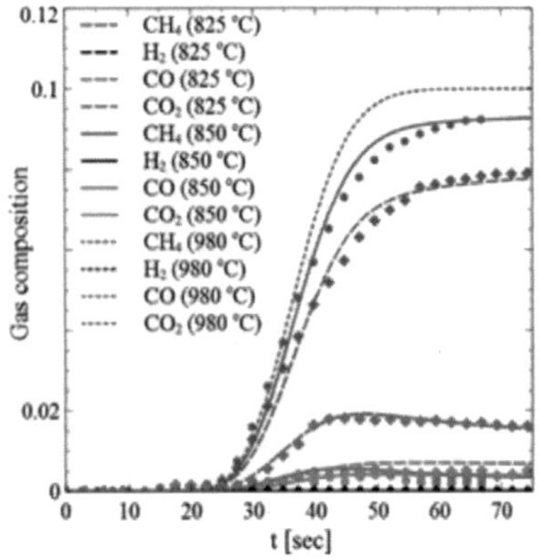

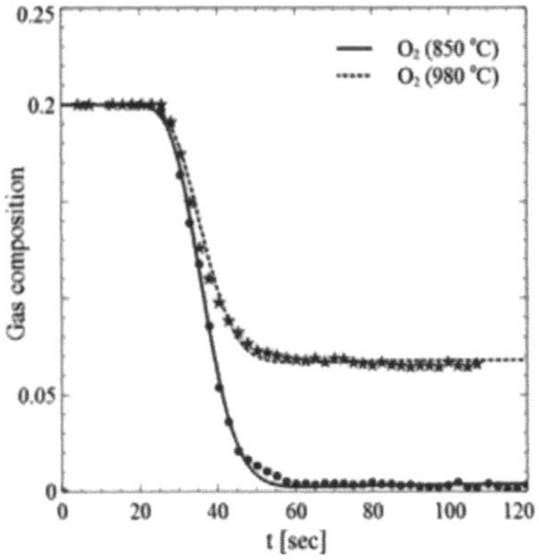

Figure 3: Model prediction and experimental data of (a) chemical-looping combustion selectivity and (b) chemical-looping with oxygen uncoupling using Cu/CuO and CH_4/Ar in a fixed-bed reactor. *Note:* model prediction at 825 °C (dashed line), 850 °C (solid line) and 980 °C (dotted line); experimental data at 825 °C (diamond), 850 °C (circle) and 980 °C (star).

Perovskite Based Oxygen Carriers for Chemical Looping with Oxygen Uncoupling (CLOU)

*Nathan Galinsky, Amit Mishra, and Fanxing Li**

Department of Chemical and Bimolecular Engineering
North Carolina State University
Raleigh, NC 27695
* Email: fli5@ncsu.edu

Chemical looping processes utilize metal oxides to store and release lattice oxygen to oxidize gaseous and solid carbonaceous fuels such as methane, syngas, coal, and biomass. To regenerate the oxygen released during the fuel oxidation step, the metal oxides, or oxygen carriers, can be regenerated by reacting with air. Typical oxygen carriers rely on Fe, Mn, Cu, Co, and/or Ni to carry lattice oxygen. An inert ceramic support is often used to enhance the activity and stability of the oxygen carrier. Perovskites have recently been shown to possess good oxygen carrying properties for different chemical looping applications including CLOU and chemical looping reforming (CLR). Such perovskite based oxygen carriers with CLOU properties allow for oxygen to be released to gas phase at high temperature, leading to improved reaction kinetics for solid fuel conversion. In this work, we investigate the effect of dopant on the CLOU and CLR characteristics of $CaMnO_3$ perovskite. A-site dopant such as Ba and B-site dopants including a number of first row transition metal oxides are investigated. While secondary metal dopants may reduce oxygen uncoupling capability of the parent perovskite by up to 50%, they are shown to be stable for over 100 redox cycles. Such effects are explained through both experiments and simulations.

In-situ X-ray absorption spectroscopy study of Na-doped and Al_2O_3-stabilized, CuO-based oxygen carriers for chemical looping combustion and chemical looping with oxygen uncoupling

Qasim IMTIAZ,[1] Paula ABDALA[2], Agnieszka KIERZKOWSKA[1], Wouter van BEEK[2], and Christoph MÜLLER[1]

[1]Laboratory of Energy Science and Engineering, ETH Zürich, Leonhardstrasse 27, 8092 Zürich, Switzerland

[2]European Synchrotron Radiation Facility (ESRF), Grenoble, France

Abstract

Chemical looping combustion (CLC) and chemical looping with oxygen uncoupling (CLOU) are emerging carbon capture and storage (CCS) technologies that are conceptually linked to oxy-fuel combustion. In CLC, hydro-carbonaceous fuels are combusted using the lattice oxygen of an oxygen carrier, typically a transition metal oxide. In the case of CLOU, the hydro-carbonaceous fuel is combusted with molecular oxygen derived from the decomposition reaction of a transition metal oxide. The CuO-Cu_2O-Cu system is very suitable for both CLC and CLOU owing to its (i) high oxygen carrying and uncoupling capacity of, respectively, 0.2 g O_2/g CuO and 0.1 g O_2/g CuO and fast kinetics for the reduction and decomposition reactions. However, CuO has one main disadvantage, *i.e.* the low Tammann temperature of CuO of 526 °C. Therefore, CuO requires stabilization by a high Tammann temperature support, *e.g.* Al_2O_3 or $MgAl_2O_4$. Al_2O_3-stabilized, CuO-based oxygen carriers have received considerable attention owing to the low cost of Al_2O_3.[i] However, Al_2O_3 can form mixed oxides with CuO, *viz.* $CuAl_2O_4$ (spinel) and $CuAlO_2$ (delafossite). Compared to the CuO-Cu_2O couple, the $CuAl_2O_4$-$CuAlO_2$ couple possesses a low oxygen equilibrium partial pressure and a slow decomposition/re-oxidation kinetics. Therefore, the formation of $CuAl_2O_4$ negatively affects the CLOU performance of Al_2O_3-stabilized, CuO-based oxygen carriers. Interestingly, the formation of the $CuAl_2O_4$ spinel can be inhibited by the incorporation of Na^+ or K^+ ions in the CuO-Al_2O_3 matrix. Here, we utilized a sol-gel technique to synthesize Na^+ promoted (1-5 wt. %) and Al_2O_3-stabilized, CuO-based oxygen carriers containing 80 wt. % CuO. The synthesized oxygen carriers were characterized using powder X-ray diffraction (XRD), X-ray absorption spectroscopy (XAS), N_2 adsorption and temperature programmed decomposition. The cyclic oxygen uncoupling capacity of the synthesized materials was evaluated in a thermo-gravimetric analyzer (TGA) at 900 °C. The oxygen uncoupling capacity of Al_2O_3-stabilized CuO decreased with cycle number owing to the slow oxidation kinetics of $CuAlO_2$. On the other hand, Na^+ promoted, CuO-based oxygen carriers showed a stable oxygen uncoupling capacity with cycle number. *In-situ* X-ray absorption spectroscopy (XAS) revealed that the presence of Na^+ ions not only inhibits the formation of $CuAl_2O_4$ but also promotes the reduction of CuO and Cu_2O. Owing to its high and stable oxygen uncoupling capacity and fast kinetcis, Na^+ doped and Al_2O_3-stabilized CuO was identified as a very promising oxygen carrier for CLC and CLOU.

---

[i] J. Adanez, A. Abad, F. Garcia-Labiano, P. Gayan, L. F. de Diego, Prog. Energy Combust. Sci. 2012, 38, 215 – 282.

375666 Optimal CO2 Capture and Utilization (CCU) Supply Chain Network for Enhanced Oil Recovery

Abstract:

More than 60% of the total anthropogenic CO_2 emissions in the U.S. are emitted from stationary sources such as power plants, iron and steel industries, cement plants, chemical plants, refineries, gas processing plants, petrochemicals, etc. While CO_2 capture and storage (CCS) [1] shows potential to reduce CO_2 emissions from these sources, several challenges must be addressed before large-scale CCS becomes a reality. One major challenge is the lack of sufficient economic drivers to deploy CO_2 capture in power and chemical industries. Although it is possible to reduce 50–80% of the current stationary emissions in the U.S. through the optimal design of a nationwide CO_2 capture, utilization and sequestration (CCUS) supply chain network [2], even the most advanced and optimized materials and processes would cost more than $35/ton of CO_2 captured and compressed [2,3]. While significant efforts [2–10] have been made to reduce the cost of CO_2 capture and compression from stationary sources, the cost of CO_2capture remains high.

To this end, utilizing anthropogenic CO_2 from stationary sources for enhanced oil recovery (CO_2-EOR) can be an influential economic driver to encourage CO_2 capture and utilization. In fact, many of the stationary CO_2 sources in the U.S. are close to oil and gas reservoirs. CO_2-EOR potentially enables incremental oil recovery up to 15% of the original oil in place (OOIP). The amount of oil that can be recovered through CO_2-EOR is currently estimated to be almost 400 billion barrels in the U.S. alone [11]. Opportunities exist to supplement and eventually replace the naturally occurring CO_2 with CO_2 from anthropogenic sources for enhanced oil recovery (CO_2-EOR).

In this work, we propose a novel CO_2 Capture and Utilization (CCU) supply chain network model to integrate nationwide, statewide or regional CO_2 capture and CO_2-EOR activities. While doing so, we consider simultaneous selection of source plants, capture technologies, capture materials, CO_2 pipelines, locations of CO_2-EOR sites, and CO_2 utilization amounts. The objective is to maximize the net profit from the CCU supply chain network, while satisfying the minimum CO_2 demands for CO_2-EOR. The net profit is obtained by including the total annual revenue minus the total annualized cost of constructing and operating the entire CCU supply chain network. The revenue from CO_2 utilization through CO_2-EOR would be generated from selling high purity CO_2 to the prospective CO_2-EOR sites, and the total CCU cost for a source is calculated by combining the investment, operating and materials costs of flue gas dehydration, CO_2 capture, compression, transportation and injection to CO_2-EOR sites. The overall network not only maximizes the profit from CCU activities, it also contributes to the overall reduction of CO_2emissions from the stationary sources.

The novel features of our CCU supply chain network model include: (1) large network of CO_2 pipelines connecting the sources, capture plants and CO_2-EOR sites, (2) selection of sources and utilization sites to maximize the overall profit from CO_2 utilization, (3) simultaneous selection of capture materials and technologies with variable CO_2 capture amounts, (4) rigorous modeling, simulation and optimization of flue gas dehydration, CO_2 capture and compression, and transportation alternatives to obtain CCU costs, (5) no CO_2 sequestration in saline formations, (6) CO_2 capture with variable CO_2 recovery, instead of considering a minimum specification of 90% on CO_2 recovery, and (7) consideration of real geographic locations of sources and utilization sites, reliable estimates of CO_2 emissions and demands. Each selected source in the CCU network supplies to one or multiple CO_2-EOR sites to satisfy different CO_2 utilization demands. The CO_2 recovery from a source can be adjusted based on the demand of nearby CO_2-EOR sites. We show that such flexibility in CO_2 recovery significantly reduces the overall CCU costs and increases the net profit. Through designing nationwide, statewide and regional CCU supply chain networks, we also quantitatively demonstrate the applicability of CO_2-EOR in reducing CO_2 emissions in an economically viable manner. The proposed network of CO_2 pipelines would also enable CO_2transportation between sources and utilization sites which are located apart from each other by as far as 200 miles.

References

[1] Metz, B.; Davidson, O.; De Coninck, H. C.; Loos, M.; Meyer, L. A. IPCC Special Report on Carbon Dioxide Capture and Storage: Prepared by Working Group III of the Intergovernmental Panel on Climate Change.IPCC; Cambridge University Press: Cambridge, United Kingdom, 2005.

[2] Hasan, M.M.F.; Boukouvala, F.; First, E.L.; Floudas, C.A. Nationwide, regional and statewide CO_2capture, utilization and sequestration supply chain network optimization. Ind. Eng. Chem. Res., 2014, 53(18), 7489 - 7506.

[3] Hasan, M.M.F.; First, E.L.; Floudas, C.A. Cost-effective CO_2 capture based on *in silico*screening of zeolites and process optimization. Phys. Chem. Chem. Phys., 2013, 15(40), 17601 - 17618.

[4] Boot-Handford, M.E.; Abanades, J.C.; Anthony, E.J.; Blunt, M.J.; Brandani, S.; Mac Dowell, N.; Fernández, J.R.; Ferrari, M.-C.; Gross, R.; Hallett, J.P.; Haszeldine, R.S.; Heptonstall, P.; Lyngfelt, A.; Makuch, Z.; Mangano, E.; Porter, R.T.J.; Pourkashanian, M.; Rochelle, G.T.; Shah, N.; Yao, J.G.; Fennell, P.S. Carbon capture and storage

update. Energy Environ. Sci., 2014, 7, 130 – 189.

[5] Hasan, M.M.F.; Baliban, R.C.; Elia, J.A.; Floudas, C.A. Modeling, simulation and optimization of post-combustion CO_2capture for variable feed concentration and flow rate. 1. Chemical absorption and membrane processes. Ind. Eng. Chem. Res., 2012, 51(48), 15642 - 15664.

[6] Hasan, M.M.F.; Baliban, R.C.; Elia, J.A.; Floudas, C.A. Modeling, simulation and optimization of post-combustion CO_2capture for variable feed concentration and flow rate. 2. Pressure swing adsorption and vacuum swing adsorption processes. Ind. Eng. Chem. Res., 2012, 51(48), 15665 - 15682.

[7] Cost and Performance Baseline for Fossil Energy Plants. Vol. 1: Bituminous Coal and natural Gas to Electricity, Revision 2, Report DOE/NETL-2010/1397; U.S. Department of Energy's National Energy Technology Laboratory: Pittsburg, PA, November 2010.

[8] DOE 2010 Report of the interagency task force on carbon capture and storage. Interagency Task Force on Carbon Capture and Storage: Washington, DC, August 2010.

[9] Ho, M. T.; Allinson, G. W.; Wiley, D. E. Reducing the cost of CO_2capture from flue gases using pressure swing adsorption. Ind. Eng. Chem.Res., 2008, 47, 4883–4890.

[10] Raksajati, A.; Ho, M.T.; Wiley, D.E. Reducing the cost of CO_2capture from flue gas using aqueous chemical absorption. Ind. Eng. Chem. Res., 2013, 52, 16887–16901.

[11] Godec, M.L.; Kuuskraa, V.A.; Dipietro, P. Opportunities for using anthropogenic CO_2 for enhanced oil recovery and CO_2 storage. Energy & Fuels, 2014, DOI: 10.1021/ef302040u.

Optimal Synthesis of CO_2 Utilization Processes Integrated with a Supercritical Pulverized Coal Power plant

Job S. Kasule[1], Debangsu Bhattacharyya[1*], Richard Turton[1], Stephen E. Zitney[2]

[1]Department of Chemical Engineering, West Virginia University, Morgantown, WV 26506

[2]U.S. Department of Energy, National Energy Technology Laboratory, Morgantown, WV 26507

Abstract

In the future, process and energy plants that produce large amounts of carbon dioxide (CO_2) are expected to have acid gas removal plants that capture a substantial amount of the CO_2 generated. A slip stream of the captured CO_2 can be utilized for producing value added products, such as chemicals, that can improve the overall plant economics and its operating efficiency. In addition, CO_2 can serve as a source of carbon thus reducing usage of other carbon sources. For the supercritical pulverized coal (SCPC) power plants where substantial waste heat is available, use of this heat for producing valuable chemicals can improve the overall plant economics significantly. With this motivation, the focus of this work is to optimally synthesize a plant with one or more CO_2 utilization processes integrated with a SCPC plant.

While CO_2 can be utilized for producing a number of chemicals, the focus of this work is mainly on producing fuels from the captured CO_2. Considering the large volume of CO_2 emitted from power plants, production of fuel seems to be a promising option without saturating the market. To this end, rigorous first-principles models of traditional as well as novel processes have been developed. In particular, models of the tri-reforming process for production of syngas, the methanol synthesis process, the photochemical reduction process for producing methanol, and the electrochemical reduction process for conversion of CO_2 to formate/formic acid have been developed. In addition, a rigorous model of a supercritical pulverized (SCPC) power plant with amine-based CO_2 capture has been developed. The SCPC plant serves as the source of CO_2 and waste energy. Not only the waste heat from the SCPC plant is utilized in CO_2 utilization processes, the waste heat from the CO_2 utilization processes are synergistically utilized in the SCPC plant especially for the CO_2 capture process. A multi-objective optimization problem for super-structure synthesis is formulated for maximizing the profit and extent of CO_2 utilization subject to a number of constraints. The resulting mixed integer nonlinear programming problem is solved by using a novel multi-software platform where the process models are developed in Aspen Plus and Aspen Custom Modeler, while the optimization layer is developed in powerful optimization software modeFRONTIER with capability to work in a multi-software environment. The presentation will include the Pareto front which captures the trade-offs between competing objectives in response to a number of disturbance variables such as prices of raw material, products, and utilities, change in the market demand, and change in the environmental emission cost.

Process and Economic Analysis of Synthetic Fuel Production from CO_2

I. Dimitriou[1*], P. Garcia-Guttierez[1], R. H. Elder[1], R. W. K. Allen[1]

[1]Department of Chemical & Biological Engineering, University of Sheffield, Mappin Street, Sheffield. S1 3JD, UK

Abstract

It is believed that the recent increase in man-made emissions of greenhouse gases, such as carbon dioxide (CO_2) is the main reason behind the observed rise in average global temperatures. Instead of emitting CO_2 into the atmosphere, utilising CO_2 as feedstock for chemicals and fuels could be a possible measure for mitigating climate change and reducing dependence on fossil fuels. For these reasons, there is increasing, world-wide interest in Carbon Dioxide Utilization (CDU) which could generate value from re-using CO_2 instead of simply disposing it underground.

Liquid fuels (e.g. methanol, hydrocarbons) can be produced via hydrogenation of CO_2 and could offer a feasible alternative for the transportation sector which is responsible for a large part of global CO_2 emissions. Unlike oxygenates, fuels produced from syngas (H_2 and CO) via Fischer-Tropsch synthesis, are compatible with conventional fuels in any proportion and thus can take advantage of existing infrastructure and supply chains. The production of hydrocarbon fuels from syngas produced from CO_2 hydrogenation processes has yet to be demonstrated. In addition, there is a noticeable lack of process feasibility studies in the literature which could assess the technical and economic potential of CO_2-to-hydrocarbon fuels routes.

This work considers three CDU process concepts for the conversion of a carbon source, in this case sewage sludge, into liquid hydrocarbon fuels. The main objective is to identify the most promising process in terms of conversion efficiencies and costs. Biogas production from anaerobic digestion of sewage sludge was chosen as the target process for its ability to act as a real world application requiring moderate capital investment. Moreover, biogas contains a significant amount of methane (approximately 65 vol%) that could be used as hydrogen source within the process, thus decreasing dependence on fossil fuels. The current study is part of the EPSRC funded 4CU research programme that examines novel technologies for carbon dioxide conversion to fuels. The process concepts examined here act as baseline scenarios that include the best currently available and proven technologies. Each subsequent development in the Programme will then be modelled against these base cases to assess the improvement that it could have on process performance and costs.

The evaluated CCU process concepts were modelled using industry standard process simulation software (Aspen Plus®) and were based on three typical biogas applications: (1) biogas upgrading using a monoethanolamine (MEA) acid gas removal unit, (2) combustion of biogas in a Combined Heat and Power (CHP) plant and (3) combustion of upgraded biogas in a CHP plant which is a combination of the first two options. All three concepts included conversion of CO_2 to carbon monoxide via a reverse water-gas-shift reaction process and subsequent conversion to diesel and gasoline via Fischer-Tropsch synthesis. The studied CCU options are compared in terms of mass yields, energy requirements, system efficiencies, capital investment and production costs. A sensitivity analysis is also carried out to examine the effect of several economic and technical parameters on the production cost of liquid fuels and results are presented.

*Corresponding author. Tel.: +44 (0) 114 222 7599; E-mail address: i.dimitriou@sheffield.ac.uk (I. Dimitriou)

Co-electrolysis of Carbon Dioxide and Steam for a Carbon Neutral Fuel Cycle

Rachael Elder, Denis Cumming, Kayode Omojola, Jen Manerova

Converting carbon dioxide to synthetic fuel has great potential for a carbon neutral fuel cycle. Co-electrolysis of carbon dioxide and steam in a high-temperature solid oxide cell is one promising method for the large-scale production of syngas; a mixture of carbon monoxide and hydrogen. Syngas is a convenient precursor for the manufacture of longer chain synthetic fuels.

Despite increasing research, one of the issues facing widespread adoption of co-electrolysis is poor understanding of the reaction mechanisms – electrolysis of carbon dioxide, electrolysis of steam and the water gas shift reaction are all possible and interact with each other. Understanding the reaction mechanisms, and their dependence on operating conditions and cell materials and morphology, is essential for commercialisation. In this paper we present extensive electrochemical measurements in combination with in-situ spectroscopic data, aimed at elucidating the processes occurring in an operating cell. Both i-V curves and electrochemical impedance spectra will be presented, along with Raman and IR spectra. Rate limiting processes are identified as well as the influence of operating temperature and flowrate.

Ultimately, understanding of the reaction mechanisms will allow the optimisation of the materials and microstructures necessary for improved device performance and large scale CO_2 recycling.

A generalized framework for the assessment of solar fuels technologies

Jeffrey A. Herron, Jiyong Kim, Aniruddha A. Upadhye, George W. Huber, Christos T. Maravelias*

Department of Chemical and Biological Engineering, University of Wisconsin – Madison, Madison, WI 53706, USA

Reducing our greenhouse gas emissions, while improving the global standard of living, is one of the key fundamental challenges of the 21st century. One of the options that has been proposed to reduce greenhouse gas emissions is the conversion of carbon dioxide and water into fuels and chemicals. Towards realizing such a solar refinery, there are many technological advances that must be met in terms of capturing and sourcing the feedstocks (namely CO_2, H_2O, and solar energy) and in catalytically converting CO_2 and H_2O.[1-5] Several approaches have been used for solar fuel production including solar driven electrochemical, photo-electrochemical and photo-catalytic CO_2 conversion to fuels, CO_2 catalytic conversion using solar-derived hydrogen (e.g. methanol synthesis for CO_2 and H_2[6]) and solar-driven thermochemical processes.

In this paper, we utilize conceptual process modeling and techno-economic analysis to assess the energy efficiency and economic feasibility of solar refineries operating through two principal CO_2 conversion routes: (1) catalytic conversion using solar-derived hydrogen, and (2) direct CO_2 reduction with H_2O. In the first route, hydrogen can be produced using solar energy through various technologies, including electrolysis, photo-electrochemical water splitting, photo-catalytic water splitting, and thermochemical methods. In the second route, CO_2 is reduced with H_2O using electrocatalytic, photo-electrochemical, photo-catalytic, or thermochemical methods.

To assess these routes and the technologies involved, we developed a process model for a generic CO_2 conversion process, which consists of five subsystems: A) CO_2 capture from source (i.e. flue gas) and transportation to solar refinery, B) CO_2 conversion using solar energy, C) vapor and liquid separation system, D) vapor product purification system, and E) liquid product purification system. For this generic process, we developed material and energy balances. Based on the material and energy balances, we developed an energy model to asses the overall energetic feasibility of the process, and we developed an economic model based on discounted cash flow analysis to calculate the minimum selling price of the product . These generic models were employed to study a specific case where *methanol is the final product* of CO_2 conversion. Finally, we performed sensitivity analysis on the specific case study to identify the key separation(s) and catalytic challenges for improving the overall economic feasibility of the solar to chemical energy process.

1. N. S. Lewis and D. G. Nocera, *Proceedings of the National Academy of Sciences*, 2006, **103**, 15729-15735.
2. G. Centi, E. A. Quadrelli and S. Perathoner, *Energy & Environmental Science*, 2013, **6**, 1711-1731.
3. A. Steinfeld, *Solar Energy*, 2005, **78**, 603-615.
4. B. A. Pinaud, J. D. Benck, L. C. Seitz, A. J. Forman, Z. B. Chen, T. G. Deutsch, B. D. James, K. N. Baum, G. N. Baum, S. Ardo, H. L. Wang, E. Miller and T. F. Jaramillo, *Energy & Environmental Science*, 2013, **6**, 1983-2002.
5. S. C. Roy, O. K. Varghese, M. Paulose and C. A. Grimes, *ACS Nano*, 2010, **4**, 1259-1278.
6. O. S. Joo, K. D. Jung, I. Moon, A. Y. Rozovskii, G. I. Lin, S. H. Han and S. J. Uhm, *Industrial & Engineering Chemistry Research*, 1999, **38**, 1808-1812.

THERMOCHEMICAL DISSOCIATION OF CO_2 INTO RENEWABLE FUELS VIA $Ce_xZr_yHf_zO_2$ BASED REDOX REACTIONS USING CONCENTRATED SOLAR ENERGY

Rahul Bhosale[1,], Ivo Alxneit[2], Anand Kumar[1], LJP van den Broeke[1], Mehak Jilani[1], Jamila Folady[1], Dareen Dardor[1], Shahd Gharbia[1], Roghayeh Dejan[1], Noor Al-Emadi[1], Fatima Al-Massih[1], Shrooq Al-yazedi[1]*

[1]Department of Chemical Engineering, College of Engineering, Qatar University, PO Box 2713, Doha, Qatar.
[2]Solar Technology Laboratory, Paul Scherrer Institute, CH-5232 Villigen PSI, Switzerland.
**Corresponding Author: rahul.bhosale@qu.edu.qa*

Abstract

In this paper we report the co-precipitation synthesis of $Ce_{0.75}Zr_{0.125}Hf_{0.125}O_2$ redox material for the production of solar fuels via thermochemical H_2O/CO_2 splitting reaction. For the synthesis of $Ce_{0.75}Zr_{0.125}Hf_{0.125}O_2$ via co-precipitation method, Ce, Zr, and Hf metal salts were dissolved in demineralized water and aqueous NH_4OH was added until the precipitation was complete. After precipitation, the slurry was filtered and the solids obtained after filtration were dried, crushed, and subsequently heated in a muffle furnace in presence of air. Calcined powders obtained were further characterized using powder x-ray diffraction (XRD), scanning electron microscopy (SEM), energy dispersive spectroscopy (EDS), and BET surface area analyzer. Solar fuel production ability and thermal stability of the synthesized $Ce_{0.75}Zr_{0.125}Hf_{0.125}O_2$ redox material was investigated by performing multiple thermochemical oxidation and thermal reduction cycles in presence of CO_2 and inert Ar using a thermogravimetric analyzer (TGA).

Introduction

Due to the incessant increase in the population of world, extreme utilization of fossil fuels led to a rapid diminution of the easily accessible oil reserves resulting in continuously rising oil prices. Apart from the fuel demands and prices, combustion of fossil fuels is believed to be one of the major causes of increase in the CO_2 concentration level in the environment. The problem of CO_2 emission is another global question concerning all societies [1-3]. The increase in CO_2 emission into the environment is of serious concern for future because CO_2 is the main reason for the greenhouse effect, which leads to global warming, glaciers melting, sea level rise, and extreme climates. Under this background, there is a pressing need to search for carbon neutral alternative energy sources which will replace fossil fuels to avoid such adverse effects of CO_2 emission.

Solar fuel production via metal oxide (MO) based thermochemical H_2O and/or CO_2 splitting reactions are considered as one of the promising new technologies for fulfillment of future energy requirement and steadying the greenhouse gas concentration in the earth's atmosphere. It is a two-step process in which the first step is related to the endothermic solar thermal reduction of MO into lower valence MO or metal and second step corresponds to the exothermic re-oxidation of reduced MO via H_2O and/or CO_2 splitting reactions. In previous investigations, two types of MO based redox systems were examined for the solar thermochemical H_2O and/or CO_2 splitting, namely: 1) volatile MO pairs such as ZnO/Zn [4-6] or SnO_2/SnO/Sn [7-8] and 2) non-volatile MO pairs such as Fe_3O_4/FeO [9-10], ferrites [11-18], and ceria based oxides [19-24]. Among all these redox systems, currently, ceria-based oxides are of particular interest of several researchers because of its faster reaction kinetics and better thermal stability as compared to other MO system. A typical ceria based two-step solar thermochemical H_2O/CO_2 splitting process is depicted in Fig. 1.

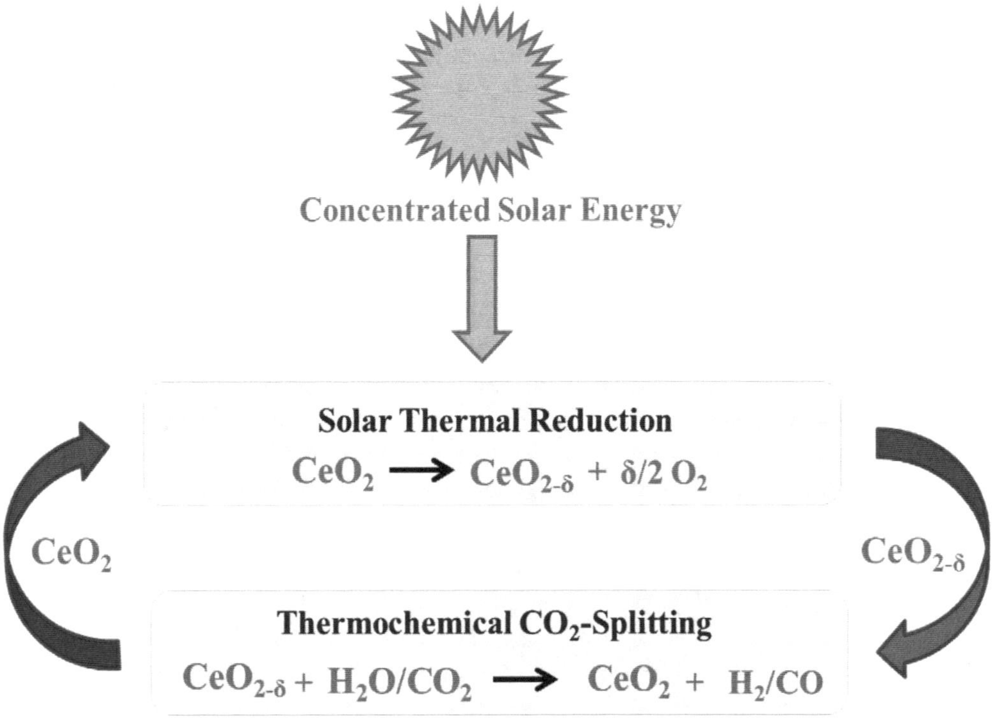

Fig.1. Typical ceria based two-step solar thermochemical H_2O/CO_2 splitting process.

Previous investigations related to utilization of ceria-based oxides for solar thermochemical fuel production reports that $Ce_xM_yO_z$ (where, M = Zr or Hf) are the best materials in terms of their thermal reduction potential, H_2O/CO_2 splitting ability and thermal stability [19-24]. However, it is important to note that, the effectiveness of ceria doped with both Zr and Hf together ($Ce_xZr_yHf_zO_{2-\delta}$) towards solar fuel production is not yet investigated. Therefore, in this study, attempts were made towards synthesizing Zr and Hf doped ceria ($Ce_{0.75}Zr_{0.125}Hf_{0.125}O_2$) using co-precipitation method and testing of its reactivity and thermal stability in multiple thermochemical CO_2 splitting cycles. $Ce_{0.75}Zr_{0.125}Hf_{0.125}O_2$ redox material, obtained via co-precipitation method and subsequent high temperature calcination in air, was characterized using powder x-ray diffraction (XRD), scanning electron microscopy (SEM), energy dispersive spectroscopy (EDS), and BET surface area analyzer. The redox reactivity of $Ce_{0.75}Zr_{0.125}Hf_{0.125}O_2$ redox material was examined by performing multiple thermochemical CO_2-Splitting cycles using a thermogravimetric analyzer (TGA).

Experimental

Synthesis of $Ce_{0.75}Zr_{0.125}Hf_{0.125}O_2$

For the synthesis of $Ce_{0.75}Zr_{0.125}Hf_{0.125}O_2$ redox material by co-precipitation method, nitrate salt of cerium, and chloride salts of zirconium and hafnium, and ammonium hydroxide (NH_4OH, 28%) were used as raw materials. As a first step, required amounts of metal salts were added and dissolved in demineralized water under mild stirring conditions. Once the metal salts are dissolved in demineralized water, NH_4OH (28%) was added slowly to the aqueous mixture containing metal salts until the precipitation was complete (pH stabilizes at equal to 9). The NH_4OH (28%) added aqueous solution containing metal salts was then stirred for 24 h and it was observed that the color of the precipitate solution changes with respect to the stirring time (Figure 2). During this stirring operation, the pH of the mixed solution was continuously monitored and additional amounts of NH_4OH (28%) was added whenever the pH value went below 9 (to maintain a constant pH = 9). The mixed solution obtained after completion of 24 h of stirring was kept undisturbed overnight and the precipitate was allowed to settle down. The precipitate was separated from the aqueous solution by using a vacuum filtration unit. During this filtration process, the precipitate was washed several times with demineralized water to get rid of the residues of synthesis. Solids obtained after filtration were dried using a vacuum furnace operated at 150°C. The dried powder obtained after vacuum drying step was further crushed using a pestle and mortar and subsequently calcined at 600°C in air using a muffle furnace.

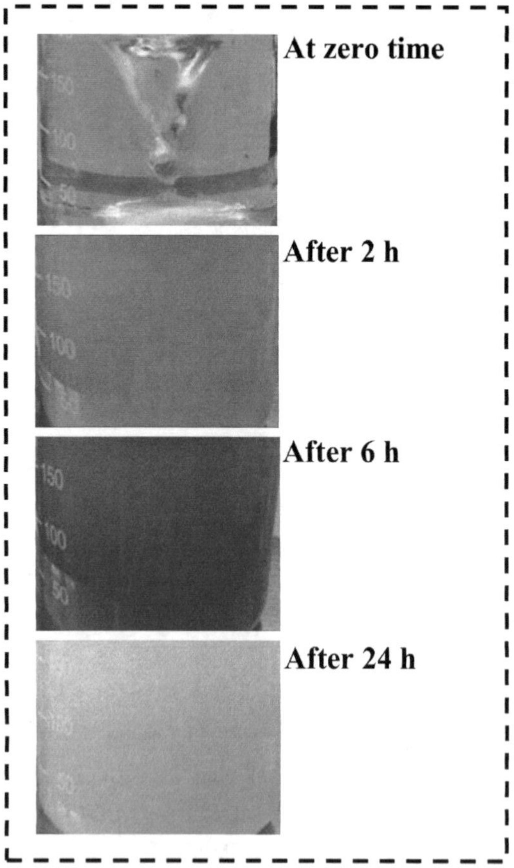

Fig. 2. Color change of $Ce_{0.75}Zr_{0.125}Hf_{0.125}O_2$ precipitate solution with respect to stirring time.

Characterization of $Ce_{0.75}Zr_{0.125}Hf_{0.125}O_2$

Phase composition, crystallite size, morphology, surface area, and elemental composition of the calcined $Ce_{0.75}Zr_{0.125}Hf_{0.125}O_2$ redox material were characterized using a Panalytical XPert MPD/DY636 powder X-ray diffractometer (XRD), a BET surface area analyzer (Micromeritics, ASAP 2420) and a Zeiss Supra 55VP field-emission scanning electron microscope (SEM) equipped with a field emission gun and X-ray energy dispersive spectrometer (EDS) from Oxford Instruments.

TGA set-up and procedure

Thermal reduction and CO_2 splitting ability of the $Ce_{0.75}Zr_{0.125}Hf_{0.125}O_2$ redox material was examined by using a thermogravimetric analyzer (TGA, Netzsch STA 409). Approximately 50 mg of calcined $Ce_{0.75}Zr_{0.125}Hf_{0.125}O_2$ powder was placed inside an Al_2O_3 crucible and was subjected to multiple thermal reduction (1400°C) and CO_2 splitting (1000°C) cycles. During the thermal reduction step, $Ce_{0.75}Zr_{0.125}Hf_{0.125}O_2$ redox material was heated upto 1400°C and maintained at this temperature for 60 min in an inert Ar atmosphere (100 ml/min). The mass loss observed after 1000°C was directly correlated towards O_2 released by the $Ce_{0.75}Zr_{0.125}Hf_{0.125}O_2$ redox material. Once the thermal reduction was carried out for 1h, the temperature was lowered to 1000°C to perform the CO_2 splitting reaction. CO_2 was injected (100 ml/min, Ar:CO_2 = 1:1) once the targeted reaction temperature was reached and the CO_2 splitting reaction was performed at isothermal conditions for 30 min. The mass gain during this step is directly correlated towards CO produced (O_2 gain by the material). Furthermore, the gases exiting the TGA were continuously analyzed via a gas chromatograph (GC, VARIAN, CP-4900, Micro GC 2 channel system).

Results and Discussion

Compositional purity of $Ce_{0.75}Zr_{0.125}Hf_{0.125}O_2$ redox material synthesized via co-precipitation method was further analyzed via powder XRD. According to the XRD patterns reported in Fig. 3a, synthesized $Ce_{0.75}Zr_{0.125}Hf_{0.125}O_2$ redox material was crystalizes with the cubic fluorite structure matching to those of pure CeO_2. Peaks associated with the dopant oxides such as ZrO_2 or HfO_2 or other impurities such as pure Zr or Hf or Ce metal were absent. It can be seen also that due to the doping of Zr and Hf in the ceria crystal structure, the XRD peak shifts to the higher 2θ angles. Furthermore, the chemical analysis of the synthesized $Ce_{0.75}Zr_{0.125}Hf_{0.125}O_2$ redox material performed via EDS confirms that the molar ratios of Ce, Zr, and Hf in the calcined powder are equal to the initially prepared aqueous solution of Ce, Zr, and Hf metal salts (Fig. 3b). The peak shift towards higher 2θ angles and the EDS analysis confirms that the $Ce_{0.75}Zr_{0.125}Hf_{0.125}O_2$ redox material was synthesized successfully via co-precipitation method.

According to the results obtained via BET analysis, it was observed that the specific surface area (SSA) of derived $Ce_{0.75}Zr_{0.125}Hf_{0.125}O_2$ redox material was observed to be equal to 38 m^2/g. It was also seen that the SSA value decreases if the calcination of the $Ce_{0.75}Zr_{0.125}Hf_{0.125}O_2$ powder obtained after filtration and drying was carried out at higher temperatures. The microstructural morphology of the $Ce_{0.75}Zr_{0.125}Hf_{0.125}O_2$ redox material was analyzed using SEM. The SEM analysis performed at higher magnification indicates spherical nanoparticles in the range of 10 to 50 nm.

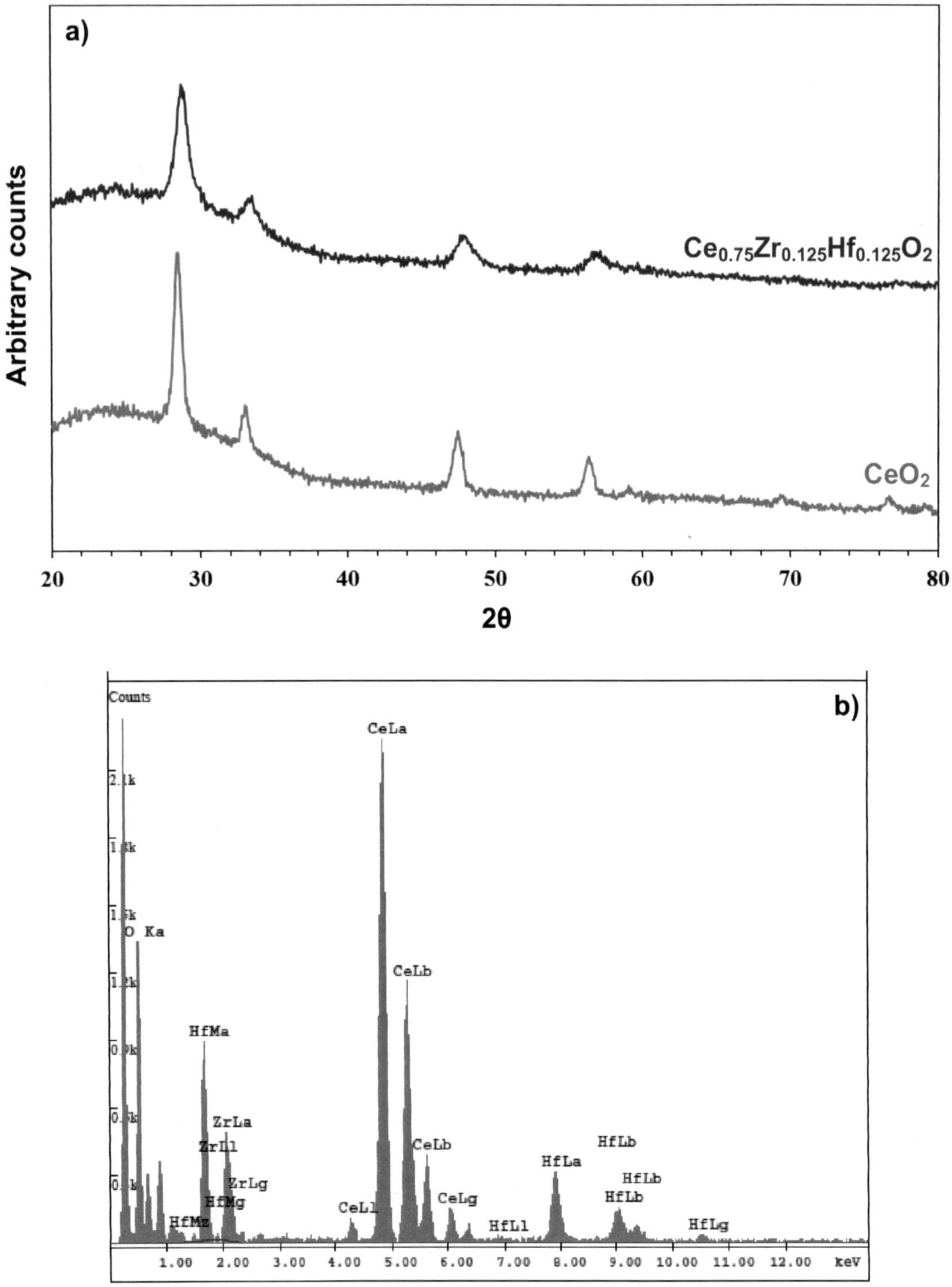

Fig. 3. a) Powder XRD patterns for CeO_2 and $Ce_{0.75}Zr_{0.125}Hf_{0.125}O_2$ redox materials, and b) EDS patterns for $Ce_{0.75}Zr_{0.125}Hf_{0.125}O_2$ redox material.

The redox reactivity of the synthesized $Ce_{0.75}Zr_{0.125}Hf_{0.125}O_2$ redox material was examined by performing multiple thermal reduction and re-oxidation via thermochemical CO_2-splitting cycles in the temperature range of 1000°C to 1400°C using a high temperature thermogravimetric analyzer. The experimental procedure is already described in the experimental section. A baseline run was conducted under identical operating conditions as in the actual experiment but in absence of a $Ce_{0.75}Zr_{0.125}Hf_{0.125}O_2$ redox material. This allows subtracting artifacts due to buoyancy effects or caused by changing the gas composition. The TGA data obtained during four thermochemical cycles performed using $Ce_{0.75}Zr_{0.125}Hf_{0.125}O_2$ redox material is presented in Fig. 4 in which temperature is reported in black color and the weight change is shown in blue color.

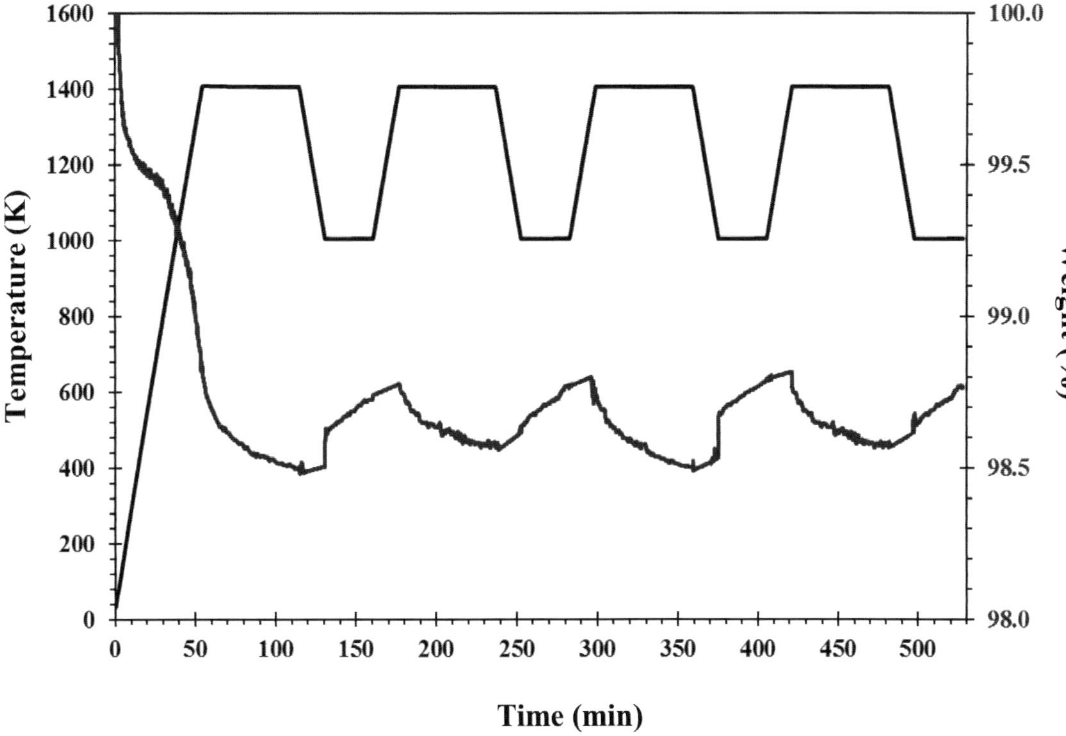

Fig. 4. TGA of CO_2-splitting experiments performed using $Ce_{0.75}Zr_{0.125}Hf_{0.125}O_2$.

According to the data reported in Fig. 4, thermal reduction of $Ce_{0.75}Zr_{0.125}Hf_{0.125}O_2$ in presence of inert Ar seems to occur at much slower rate than the re-oxidation via thermochemical CO_2 splitting reaction. During the four cycles presented, the redox capacity of $Ce_{0.75}Zr_{0.125}Hf_{0.125}O_2$ does not deteriorate during 2nd, 3rd, and 4th thermal reduction and re-oxidation step. However, it is noteworthy that during the first reduction a disproportionally large weight loss occurs. We attribute a significant fraction of this weight loss to the desorption of physisorbed water from the samples. Amounts of O_2 released and CO produced in consecutive thermochemical cycles performed using $Ce_{0.75}Zr_{0.125}Hf_{0.125}O_2$ redox material were estimated with the help of % weight changes (TGA analysis) and the data obtained via gas chromatography. During the first thermal reduction and CO_2-splitting cycle, 220 μmol/g of O_2 and 125 μmol/g of CO produced. However, in the 2nd, 3rd, and 4th thermal reduction and CO_2-

splitting cycles, an average of 75 μmol/g·cycle O_2 and 105 μmol/g·cycle of CO were produced with CO to O_2 ratio equal to 1.40.

In previous investigation [23], it was reported that $Ce_{0.75}Zr_{0.25}O_2$ produced 92 μmol/g·cycle O_2 and 46 μmol/g·cycle of CO with CO to O_2 molar ratio equal to 0.5 during first three successive thermochemical CO_2 splitting and thermal reduction cycles performed at 1050°C and 1400°C, respectively. In this investigation, it was observed that the $Ce_{0.75}Zr_{0.125}Hf_{0.125}O_2$ redox material produced 110 μmol/g·cycle O_2 and 110 μmol/g·cycle of CO with CO to O_2 molar ratio equal to 1.0 during first four consecutive thermochemical CO_2 splitting and thermal reduction cycles performed at 1000°C and 1400°C. It can be seen that the amounts of CO produced and ratio of molar production of CO to O_2 is high in case of $Ce_{0.75}Zr_{0.125}Hf_{0.125}O_2$ as compared to $Ce_{0.75}Zr_{0.25}O_2$.

Conclusions

In this paper, synthesis of $Ce_{0.75}Zr_{0.125}Hf_{0.125}O_2$ redox material and its utilization towards solar thermochemical fuel production is explored. The synthesis of $Ce_{0.75}Zr_{0.125}Hf_{0.125}O_2$ redox material was achieved using co-precipitation method. The physico-chemical characterization of the synthesized $Ce_{0.75}Zr_{0.125}Hf_{0.125}O_2$ redox material was carried out by using different analytical techniques such powder XRD, SEM, BET, and EDS analysis. Derived $Ce_{0.75}Zr_{0.125}Hf_{0.125}O_2$ redox material was further examined towards its thermal reduction and re-oxidation ability in multiple thermochemical CO_2 splitting cycles using a thermogravimetric analyzer in the temperature range of 1000°C to 1400°C. Results indicate that the synthesized $Ce_{0.75}Zr_{0.125}Hf_{0.125}O_2$ redox material produced higher amounts of CO during multiple thermochemical CO2 splitting cycles with better CO to O_2 molar ratio as compared to $Ce_{0.75}Zr_{0.25}O_2$. For instance, 110 μmol/g·cycle O_2 and 110 μmol/g·cycle of CO with CO to O_2 molar ratio equal to 1.0 was generated during first four consecutive thermochemical CO_2 splitting cycles performed using $Ce_{0.75}Zr_{0.125}Hf_{0.125}O_2$. Furthermore, the redox reactivity of this material was observed to be constant during multiple thermochemical cycles.

Acknowledgments

The authors gratefully acknowledge the financial support provided by the Qatar University (QUUG-CENG-CHE-13/14-4), Indo-Swiss Joint Research Program (ISJRP, grant #138852), and Swiss Federal Office of Energy.

References

1. Bhosale, R., and Mahajani, V. (2013), "Kinetics of absorption of carbon dioxide in aqueous solution of ethylaminoethanol modified with N-methyl-2-pyrolidone", *Separation Science & Technology*, **48**, 2324.
2. Bhosale, R., and Mahajani, V. (2013), "Kinetics of thermal degradation of renewably prepared amines useful for flue gas treatment", *Journal of Renewable and Sustainable Energy*, **5**, 063110-1.
3. Bhosale, R., Gharbia, S., Folady, J., Jilani, Mehak., Dardor, D., Kumar, A., van den Broeke, L. J. P., Mahajani, V., and Dash, S. K. (2014), "CO_2 capture using an aqueous formulated solvent containing ethylaminoethanol, N-Methyl-2-Pyrolidone, and hydroxyl

radical scavengers: study of solvent degradation and absorption kinetics", *Proceedings of 4th Annual Gas Processing Symposium,* (Accepted – In Press).

4. Loutzenhiser, P., and Steinfeld, A. (2011), "Solar syngas production from CO_2 and H_2O in a two-step thermochemical cycle via Zn/ZnO redox reactions: Thermodynamic cycle analysis", *International Journal of Hydrogen Energy*, **36**, 12141.
5. Stamatiou, A., Loutzenhiser, P., and Steinfeld, A. (2010), "Solar syngas production from H_2O and CO_2 via two-step thermochemical cycles based on Zn/ZnO and FeO/Fe_3O4 redox reactions: Kinetic analysis", *Energy & Fuels*, **24**, 2716.
6. Loutzenhiser, P., Meier, A., and Steinfeld, A. (2010), "Review of the two-step H_2O/CO_2-splitting solar thermochemical cycle based on Zn/ZnO redox reactions", *Materials*, **3**, 4922.
7. Charvin, P., Abanades, F., Lemont, F., and Flamant, G. (2008), "Experimental study of SnO_2/SnO/Sn thermochemical systems for solar production of hydrogen", *AIChE Journal*, **54**, 2759.
8. Abanades, F. (2012), "CO_2 and H_2O reduction by solar thermochemical looping using SnO_2/SnO redox reactions: Thermogravimetric analysis", *International Journal of Hydrogen Energy*, **37**, 8223.
9. Abanades, F., and Villafan-Vidales, H. (2011), "CO_2 and H_2O conversion to solar fuels via two-step solar thermochemical looping using iron oxide redox pair", *Chemical Engineering Journal*, 175, 368.
10. Agrafiotis, C., Roeb, M., Konstandopoulos, A., Nalbandian, L., Zaspalis, V., Sattler, C., Stobbe, P., and Steele, A. (2005), "Solar water splitting for hydrogen production with monolithic reactors", *Solar Energy*, **79**, 409.
11. Scheffe, J., Li, J., and Weimer, A. (2010), "A spinel ferrite/hercynite water-splitting redox cycle", *International Journal of Hydrogen Energy*, **35**, 3333.
12. Bhosale, R., Shende, R., and Puszynski, J. (2010), "H_2 generation from thermochemical water splitting using sol-gel derived Ni-ferrite", *Journal of Energy and Power Engineering*, **4**, 27.
13. Bhosale, R., Shende, R., and Puszynski, J. (2010), "H_2 generation from thermochemical water-splitting using sol-gel synthesized Zn/Sn/Mn-doped Ni-ferrite", *International Review of Chemical Engineering*, **2**, 852.
14. Bhosale, R., Shende, R., and Puszynski, J. (2012), "Thermochemical water-splitting for H_2 generation using sol-gel derived Mn-ferrite in a packed bed reactor", *International Journal of Hydrogen Energy*, **37**, 8223.
15. Bhosale, R., Khadka, R., Shende, R., and Puszynski, J. (2011), "H_2 generation from two-step thermochemical water-splitting reaction using sol-gel derived $Sn_xFe_yO_z$", *Journal of Renewable and Sustainable Energy*, **3**, 063104-1.
16. Bhosale, R., Shende, R., and Puszynski, J. (2012), "Sol-gel derived $NiFe_2O_4$ modified with ZrO_2 for hydrogen generation from solar thermochemical water-splitting reaction", *Proceedings of Materials Research Society Symposium*, 1387.
17. Bhosale, R., Alxneit, I, van den Broeke, Leo L. P., Kumar, A., Jilani, M., Gharbia, S., Folady, J., and Dardor, D. (2014), "Sol-gel synthesis of nanocrystalline Ni-Ferrite and Co-Ferrite redox materials for thermochemical production of solar fuels", *Proceedings of Materials Research Society Symposium* (Accepted – In Press).

18. Fresno, F., Yoshida, T., Gokon, N., Fernandez-Saavedra, R., and Kodama, T. (2010), "Comparative study of the activity of nickel ferrites for solar hydrogen production by two-step thermochemical cycles", *International Journal of Hydrogen Energy*, **35**, 8503.
19. Abanades, S., and Flamant, G. (2006), "Thermochemical hydrogen production from a two-step solar-driven water-splitting cycle based on cerium oxides", *Solar Energy*, **80**, 1611.
20. Scheffe, J., Jacot, R., Patzke, G., and Steinfeld, A. (2013), "Synthesis, characterization and thermochemical redox performance of Hf, Zr and Sc doped ceria for splitting CO_2", *Journal of Physical Chemistry C*, **177**, 24104.
21. Chueh, W. C., Falter, C., Abbott, M., Scipio, D., Furler, P., Haile, S., and Steinfeld, A. (2010), "High-flux solar-driven thermochemical dissociation of CO_2 and H_2O using nonstoichiometric ceria", *Science*, **330**, 1797.
22. Le Gal, A., and Abanades, S. (2011), "Catalytic investigation of ceria-zirconia solid solutions for solar hydrogen production", *International Journal of Hydrogen Energy*, **36**, 4739.
23. Abanades, S., and Gal, A. L. (2012), "CO_2 splitting by thermo-chemical looping based on $Zr_xCe_{1-x}O_2$ oxygen carriers for synthetic fuel generation", *Fuel*, **102**, 180.
24. Le Gal, A., Abanades, S., and Flamant, G. (2011), "CO_2 and H_2O splitting for thermochemical production of solar fuels using nonstoichiometric ceria and ceria/zirconia solid solutions", *Energy & Fuels*, **25**, 4836.

381812 Hydrogen and Carbon Monoxide Production By Co-Electrolysis of H2O and CO2 Using Solid Oxide Cells

Abstract:

Hydrogen is currently considered as one of the most interesting energy carrier because of limited reserves of fossil fuels, the increasing price of oil and environmental regulations concerning the emission of green-house gases.However, conventional hydrogen production routes such as steam reforming, partial oxidation of heavy hydrocarbons and coal gasification requires the use of fossil fuels.Alternative processes for hydrogen production include reforming and pyrolysis using biomass, direct methanol reforming, fermentation of biomass and biological production. On the other hand, production of hydrogen and synthetic hydrocarbons from renewable energy sources is a very attractive solution to reduce both oil consumption and carbon dioxide emissions without the need for modifications of existing infrastructures.Renewable energy resources such as solar energy, wind power, hydropower or geothermal power have gained considerable attention in recent years due to their large availability and the possibility to use the excess energy for an efficient production of fuels by high temperature steam electrolysis (HTSE). The hydrogen, produced by HTSE, can be stored and directly used as fuel for combined heat and power generation (CHP) by power systems based on solid oxide fuel cells (SOFC). Reversible systems based on solid oxide cells can be integrated with renewable sources to produce gaseous fuel by using solid oxide electrolysis (SOE) that can be reused to convert the produced fuel into electricity when the energy source is not available. It is worth noticing that, among all methods to produce hydrogen, water electrolysis combined with renewable sources can be considered the cleanest route to produce electricity since no hydrocarbons are consumed and no CO_2 is released in the whole process. However, while hydrogen production by electrolysis has been under development in the last twenty years, only few studies on production of synthesis gas (H_2+CO) by simultaneous electrolysis of steam and carbon dioxide (coelectrolysis) has been reported in the last few years. Synthesis gas is the raw gas mixture for production of liquid hydrocarbon fuels via the Fischer Tropsch synthesis (FTS) and it is conventionally produced by coal gasification or steam reforming of natural gas. Nevertheless, these conventional methods for syngas production also release significant amount of CO_2 needs to be separated before feeding the mixture to the FTS reactor. Coelectrolysis is a very attractive technology for the production of syngas from renewable energy sources without consuming fossil fuels or releasing green-house gases. Conversion of CO_2 into precursors for FTS can be considered as a clean process suitable to sustain a closed carbon cycle. In addition, after removing H_2O and CO_2, the H_2/CO mixture obtained from coelectrolysis is free from gas impurities such as tar, aromatics and organic impurities which are, on the other hand, typical of syngas derived from biomasses.Coelectrolysis in SOC has only been demonstrated in few studies and there is not a general agreement whether CO is exclusively produced via the reverse water gas shift reaction and that no electrolysis of CO_2 occurs or if CO is produced both via the reverse water gas shift reaction and via electrolysis of CO_2.The total amount of Hydrogen downstream the cell depends on the steam conversion (which is proportional to the current applied to the system) but also on the H_2 percentage which is normally included in the feed gas in order to maintain reducing conditions. The composition of the syngas which is produced by coelectrolysis and therefore the ratio between H_2 and CO downstream the system should be tailored depending on the structure of the liquid fuel to be produced by FTs. A H_2 to CO molar ratio between 2 and 3 is normally required to produce light hydrocarbons.The present work is focused on production of H_2 and H_2/CO-rich mixtures by electrolysis and coelectrolysis in SOC-based stacks. The stack consists of an assembly of 6 Ni-YSZ supported cells interfaced with proprietary SOFCONNEX™ gas diffusion layers and Crofer 22 APU metallic interconnectors. The latter are coated on the air side to reduce Cr evaporation.The aim of the study is to investigate the suitability of this stack to operate in electrolysis and coelectrolysis mode and also to run under reversible SOE/SOFC operation.Measurements were performed on a proprietary test bench designed to test stacks in the power range of 20-200W.The performances of the stack were initially investigated in SOFC-mode, performing current-potential polarization curves in the range 650°C - 750°C. After that, stack operation was reversed in SOE-mode and the stack was tested under 90%H_2O/10%H_2 and 50%H_2O/10%H_2 cathode atmosphere. In order to assess the suitability of the stack to be integrated with renewable energy sources, fast on/off load cycles as well as SOFC/SOE cycles were performed and a very good performance stability was observed.Co-electrolysis measurements were performed at 750°C and using gas mixtures containing different CO_2/(H_2+H_2O) ratios. The composition of the gases, downstream the stack and after H_2O condensation, was analysed by gas chromatography (GC) in order to investigate the influence of the operating parameters and fuel composition on the relative amount of H_2, CO and CO_2 produced in the stack. As expected, the amount of H_2 in the mixture linearly increase with the increase of current density. A generally similar trend was also observed for CO percentage, while CO_2 percentage rapidly decreases with the increase of the electrolysis current. The rate of CO production was observed to depend on the relative CO_2 amount in the gas mixture entering in the stack. Thermochemical calculations were also performed in order to correlate the results of the GC analysis to the composition of the cathode gas atmosphere.

CO_2 Utilization as Oxidant for High Value Hydrocarbon

Abhishek Burri, Yong-Hwan Mo, Ranjit Kulkarni and Sang-Eon Park*

Laboratory of Nano-Green Catalysis, Department of Chemistry and Chemical Engineering, Inha University, Incheon 402-751, Republic of Korea.

Abstract

Using carbon dioxide as a feedstock chemical could help to reduce levels of this greenhouse gas in the atmosphere, as well as providing a cheap source of carbon or CO and active oxygen. The transformation of CO_2 into more useful organic substances is a promising long term objective because of the non-reactive nature of it. Carbon dioxide is a totally renewable feedstock compared to oil or coal. Carbon dioxide is generally considered to be a green or environmentally benign solvent or medium in that it is relatively nontoxic, nonflammable, and naturally abundant. Several prominent features of CO_2 have also been pointed out which include, for example, CO_2 accelerates the reaction rate, enhances the selectivity, alleviates the chemical equilibrium, suppresses the unwanted total oxidation products, prevents the hot spots on the catalyst surface, and poisons the nonselective sites of the catalysts[1]. Thus, utilization of CO_2 in oxidative dehydrogenation (ODH) of alkanes and alkylaromatics (ethylbenzene, substituted benzenes) to high value hydrocarbon conversions is very promising.

ODH of alkylaromatics and alkanes has been of interest for researchers and industrialists. Many kinds of catalysts have been developed based on the acid-base bi-functionality, redox and supported materials. Among them amphoteric ZrO_2 has been studied widely because of its robust properties, acid base bi-functionality and for its CO_2 activation capability [2] which showed positive effect in ODH of ethylbenzene to styrene with CO_2 compared to inert gas (He). Utilization of CO_2 in ODH enhanced the catalytic stability by coupling with reverse water gas shift reaction by alleviating the formed hydrogen and forwarding the reaction and also decreasing the surface coke by reverse boudouard reaction. Later, ZrO_2 along with other mixed metal oxides like TiO_2, MnO_2, CeO_2 and SnO_2 was studied and positive effects were observed. Improving the surface basicity by alkali metal has enhanced the activity, selectivity and stability at lower temperatures [3]. Further, doping of alkali into the TiO_2-ZrO_2 mixed metal oxide system gave improved catalytic activity, high surface oxygen and longevity of the catalyst [4]. This modification in doping also enhanced the CO_2 conversions which implies the ODH process enhancement compared to the thermodynamic conversion of ethylbenzene on the styrene.

References

1) M. B. Ansari and S.-E. Park., *Energy Environ. Sci., 2012, 5, 9419–9437*

2) J.-N. Park, J. Noh, J.-S. Chang and S.-E. Park, *Catal. Lett.*, 2000, 65, 75-78.

3) D. R. Burri, K.-M. Choi, S.-C. Han, A. Burri and S.-E. Park, *J. Mol. Catal. A: Chem.*, 2007, 269, 58-63.

4) A. Burri, N. Jiang and S.-E. Park, *Catal. Sci. Technol., 2012,2, 514–520.*

371654 High Temperature Mercury Oxidation Catalyst for Mercury Emissions Control

Abstract:

The use of different coals and different air pollution control device configurations within coal-fired power plants pose a great challenge for cost-effective mercury technology solutions to meet the new impending Mercury and Air Toxics Standards (MATS) rule. This study aims to develop Hg(0) oxidation catalysts that can work at typical SCR unit temperatures of 350-400 °C. Selected mercury oxidation catalysts prepared onto TiO_2 were tested in both a lab-scale fixed-bed reactor and a bench-scale reactor with structured catalysts at 370 °C. Two different simulated flue gas conditions representing eastern bituminous coal and PRB coal were used. Both fresh and spent catalysts were characterized by SEM-EDX, XAFS, FTIR, and XPS.

Under high HCl conditions, 95% Hg(0) oxidation could be achieved while minimizing metal loadings. The addition of cerium also improved the performance under high SO_2 conditions (up to 2000 ppm). For low chlorine-containing PRB and lignite coals, the injection of $CaBr_2$ to the boiler is a viable solution. The catalysts were also optimized to work with low bromine content (less than 5 ppm in the form of HBr) and the amount of $CaBr_2$ needed could be significantly reduced. Any significant SO_3 or NO_x gas generation was not observed over the catalysts. The mercury oxidation catalysts can be installed at the tail section of the existing SCR unit. Oxidized mercury in the form of $HgCl_2$ or $HgBr_2$ is highly soluble in water and can be captured in the subsequent wet FGD process.

389986 Mercury Control Strategies in Coal-Fired Power Plants

Abstract:

Mercury control in coal-fired power plants can be accomplished using two main strategies, co-benefits approach or mercury-specific approaches, and this talk will present an overview of these stratgies and how various technologies fit into the strategies. We will review specific pilot- and full-scale experimental results. Finally, we will focus on the development of future technologies in mercury control.

Co-benefits approaches generally focus on the oxidation of mercury and the dissolution of the oxidized mercury into existing wet flue gas desulfurization (WFGD) equipment. This oxidation can be accomplished by the native halides in the fuel, by the addition of halides or other oxidizing agents to combustion process, or by the co-benefit of oxidation across existing catalyst, which would be used for NOx reduction. Once oxidized, the mercury salts can be dissolved into the WFGD. However, depending upon the WFGD unit chemistry, reduction of mercury salts back to the elemental form may occur, necessitating the use of a mercury-specific approach (such as sorbent or chelating agent in the WFGD). Mercury-specific approaches generally focus on the use of sorbents for the removal of mercury, and the mercury-laden sorbents are then removed in existing electrostatic precipitator (ESP) or fabric filter (FF) baghouse. Typically the sorbents have an activated carbon base material, though non-carbon oxides, metals, and other materials have been tested over the years. There are some challenges related to these sorbents, particularly in flue gases derived from high S fuels (high SO3) where their sorption capacity is decreased. This talk will focus also on some strategies that have been used to increase capacity in high SO3 environments. Other issues include sorbent in fly ash, as well as the generally low dynamic sorption of these materials in the flue gas environment.

Several research programs, including development/testing of SO3 tolerant and non-carbon sorbents as well as control of mercury re-emission from the WFGD, have been recently performed at Southern Company. This talk will highlight those results which have been achieved at the pilot- and full-scale.
Finally, with the US EPA enforcement of Mercury and Air Toxics Standards Rule imminent, this talk will discuss the future of mercury control technology development.

384391 Evaluation of Oxidized Mercury Species Desorbed from Hg Sorbents for Coal-Fired Flue Gas Applications

Abstract:

Mercury exists in various forms in the flue gases of coal combustion, from its elemental state (Hg^0) at the boiler exit, to various oxidized forms (Hg^{2+}) as it interacts with fly ash particle surfaces as the flue gas cools. The exact species of the oxidized mercury is, at this point, an outstanding question. Thermodynamic calculations suggest that the favored form in coal combustion flue gas is $HgCl_2$, but HgO and $HgBr_2$ could exist under certain conditions, such as the introduction of Br into the combustion process or in cases when coal is burned and contains trace amounts of Br.

An electron ionization quadrupole mass spectrometer (EI-QMS) has been modified for the direct measurement of elemental and oxidized Hg at industrial relevant concentrations ($< 15\ \mu g\ m^{-3}$ or 2 ppbv) from a methane combustion flue gas. Hg^0 and HCl, Cl_2 or Br_2 are introduced pre-combustion and passed through the methane combustion flame to obtain halogen radicals, representative of the real oxidizing environment of a coal-fired boiler. Each halogen species is investigated to determine its effectiveness at oxidizing Hg^0. By burning Hg^0 with the halogen allows for the effect of homogeneous oxidation to be taken into account. Although homogeneous oxidation is expected to comprise less than 10% of the total oxidation mechanism, it is important to include this component to provide an accurate representation of the mercury species postcombustion and entering the packed-bed reactor for sorbent testing. Ultimately, the extent of homogeneous and heterogeneous oxidation may be measured, with the Hg species directly identified through mass spectrometry. In addition, the oxidized species created upon exposure to surfaces present in coal combustion, such as fly ash and activated carbon, are identified through the use of packed-bed experiments. A variety of sorbents have been evaluated, with the MS used to track any desorbed species. This information is critical for evaluating the extent of reaction on a sorbent surface and the potential for desorption of various oxidized Hg species. By investigating the desorbed species as a function of temperature provides indication of the relative strength of adsorption of the various Hg species and subsequently their relative stabilities on the various sorbents tested. These types of investigations will also increase the fundamental understanding of the surface reaction of Hg on a variety of sorbents. A commercial Hg analyzer and speciation system is used to benchmark total oxidized Hg species determined by mass spectrometry since this is the first time Hg speciation of the oxidized have been directly measured and at realistic concentrations.

Sravan Pappu, Andrea Foster, Rachel Mansfield
Johnson Matthey Process Technologies

Title: Hg Removal Technologies For Hydrocarbon, Reducing, and Water Streams

This paper aims to review the current status of Hg removal in refineries and potentially other industrial processes. Topics that will be included are the regulatory requirements and other drivers for Hg removal, current technology solutions, and novel advances in Hg removal from hydrocarbon, reducing, and water streams. Focus will be given to recent advances in removal technology and how refineries and other industrial users can address challenges they may face with mercury removal.

Mercury removal from hydrocarbon streams within the refinery, as heavy as naphtha, is a well-established process using a fixed bed of sorbents. Operators in this area are looking to achieve long bed lives with decreased frequency between changeouts. This reduces the environmental, health, and safety impacts associated with frequent material replacement and maintains the Hg removal required for downstream processes and product specifications. Operators are also interested in low pressure drop across the unit during operation. These requirements require high Hg pickup and robust physical characteristics. New advances have enabled a product that allows refiners to minimize maintenance activity and ensure reliable Hg removal even when faced with certain process disturbances.

The second application to be discussed is in Hg removal from reducing environments. In some industrial processes, vent streams are generated that contain hydrogen. Conventional mercury removal beds are either unsuitable for these streams or do not offer removal to low levels. A new absorbent has been developed that gives both high capacity and low exit mercury in these duties.

Finally, Hg removal from water will potentially see increased scrutiny from regulatory agencies, however measuring / removing such content has not been effectively addressed for refineries and other industrial operators. Such requirements can cause significant impact to businesses not ready to approach Hg content with solutions that are fit for purpose. This discussion will focus on a newly available solution, its removal mechanism, and performance data.

389647 Elemental Mercury Oxidation in an Electrostatic Precipitator with in-Situ Soft X-Rays

Abstract:

Nonthermal plasma--based techniques are considered promising for oxidizing elemental mercury (Hg^0) in flue gas from coal combustion. In this study, we investigated Hg^0 oxidation in a DC (direct current) corona electrostatic precipitator (ESP) coupled with in-situ soft X-rays. Soft X-rays enhanced Hg^0 oxidation by promoting electron generation, creating a large quantity of oxidizing radicals, such as monoatomic oxygen (O) and ozone molecules (O_3). Positive corona discharges worked better in Hg^0 oxidation than negative ones, because they generate more high energy electrons, a finding which was supported by plasma simulations of soft X-ray--coupled ESP. Reaction mechanisms were further studied in single gas environments where pure oxygen (O_2), nitrogen (N_2), or carbon dioxide (CO_2) was the carrier gas. In the CO_2 case, Hg^0 could not be oxidized by ESP operation alone, but it could be oxidized with ESP and soft X-rays. This result indicates that soft X-ray enhancement has even greater potential for Hg^0 oxidation by ESP when high concentrations of CO_2 are present in flue gas, such as in the oxy-combustion case. The effects of titanium dioxide (TiO_2), silicon dioxide (SiO_2), and potassium iodide (KI) particles on the oxidation processes were also studied. When the soft X-ray--coupled ESP technology is applied to mercury capture from flue gas, it still retains the co-benefit of capturing particles.

Mercury (Hg) is a global pollutant that affects human and ecosystem health. In 1997 the US EPA completed a comprehensive Mercury Study Report to Congress, identifying mercury as the hazardous air pollutant of greatest potential public health concern. Recent studies indicate that about 70-85% of total anthropogenic Hg emissions are caused by combustion sources, mainly coal-fired utilities and waste incinerators. Although multi-pollutants' reductions from flue gas using different processes, e.g.; NO_X, SO_2, and Hg removal by Selective catalytic reduction (SCR), lime/limestone wet flue gas desulfurization (WFGD) and activated carbon, respectively, are efficient purification techniques, they are involved with high construction and operational cost, as well as large installation space for the complex treatment processes. WFGDs known for their high removal efficiency, reasonable cost and wide applicability, have been shown, in general, to achieve less than 70% mercury removal efficiency. However, based on preliminary economic estimations, flue gas cleaning plants with a multifunctional scrubber have lower investment costs than state-of-the-art systems. With the number of WFGD scrubbers likely to be constructed in response to the Clean Air Transport Rule (CAIR) issued by EPA on July, 6, 2011, it is desirable to develop cost-effective, wet-scrubber-based technologies, especially those capable of providing simultaneous multipollutant (SO_2, NO_X, and Hg^0 and Hg^{2+}) control and could be possibly retrofitted to existing power plant. Such technologies could make wet-FGD scrubber more cost-effective and could obviate the need for installation of additional costly control equipment such as SCR system.

Certain strong aqueous chemical oxidants can be effectively, benignly and economically used to reduce $Hg^0/NO_x/SO_2$ emission simultaneously from stationary sources with a single wet scrubber. We recently published several studies on NO_X (and simultaneous NO_X/SO_2) absorption and oxidation by persulfate activated by temperature with/without Fe(II) ions and EDTA. The ultimate goal of this study is to develop an innovative and cost-effective chemical scrubber for the simultaneous removal of mercury, nitric oxide and sulfur dioxide from flue gases under WFGD scrubber operating conditions using aqueous solutions of sodium persulfate ($Na_2S_2O_8$). In this talk, a comprehensive review of the chemistry and mechanisms of Hg^0 by strong chemical oxidants in aqueous scrubbing systems in the presence and absence of NO_X, SO_2 and other flue gas components, and our current work using aqueous solutions of sodium persulfate ($Na_2S_2O_8$) will be discussed.

References:
Adewuyi, Y.G.; Khan, M.A.; Sakyi, N.Y. Ind. Eng. Chem. Res. 2014, 53, 828.
Adewuyi, Y.G.; Sakyi, N.Y. Ind. Eng. Chem. Res. 2013, 52, 11702.
Adewuyi, Y.G.; Sakyi, N.Y. Ind. Eng. Chem. Res. 2013, 52, 14687.
Khan, N.E.; Adewuyi, Y.G. Ind. Eng. Chem. Res. 2010, 49, 8749.

2014 AIChE Annual Meeting

Rapid Detection of Trace Element Concentrations Using Laser Induced Breakdown Spectroscopy and Laser-Breakdown Time-of-Flight Mass Spectrometry

Yoshihiro Deguchi

The University of Tokushima

Outline

1. **Introduction**
2. **Theory**
3. **Measurement of trace species using low pressure LIBS**
4. **Measurement of trace species using LB-TOFMS**
5. **Conclusions**

LIBS: Laser-induced breakdown spectroscopy
LB-TOFMS: Laser breakdown time-of-flight mass spectrometry

1. Introduction

Energy issue

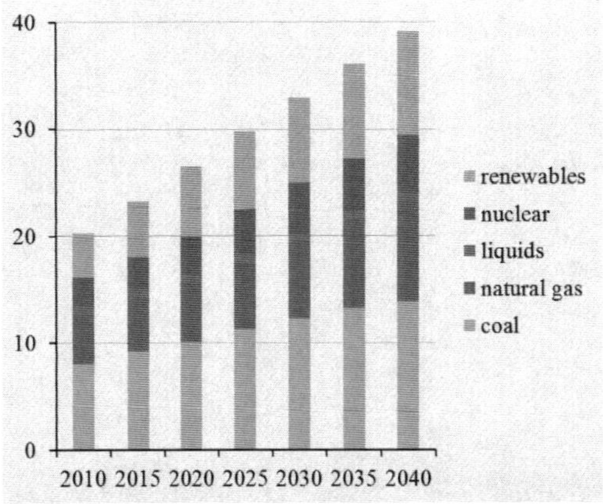

net electricity generation by energy source

Source: International energy outlook 2013-EIA

Thermal power plant

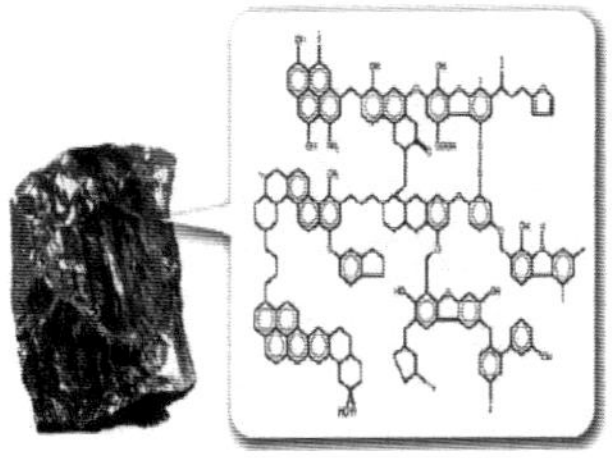

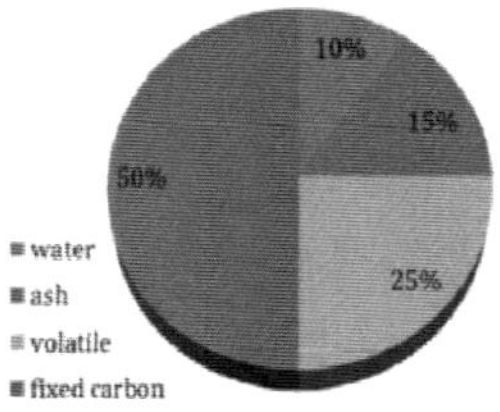

Type of coal

Type	Calorific value (Kcal/kg)	Fuel ratio	Cohesive property
anthracite	—	>4.0	non-cohesive
bituminous coal	8400	>1.5 <1.5	strong-cohesive
	8100	>1.0 <1.0	cohesive weak-cohesive
semi bituminous coal	7800	>1.0	weak-cohesive
		<1.0	non-cohesive
	7300	—	non-cohesive
lignite	6800 5800	—	non-cohesive

1. Introduction

Research items

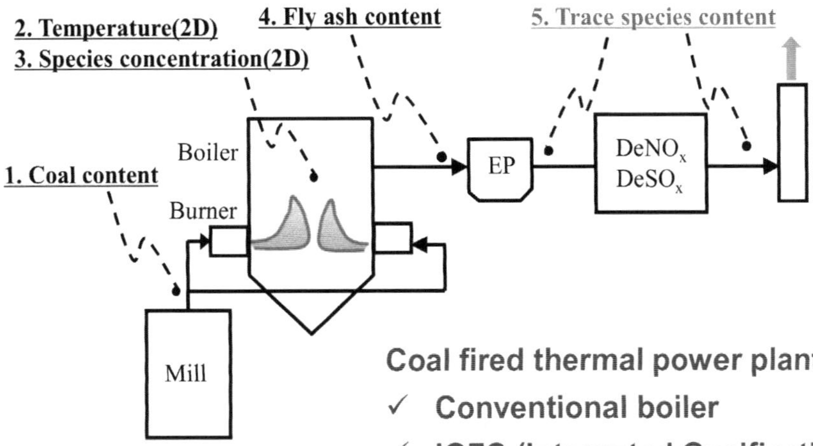

Coal fired thermal power plants

- ✓ Conventional boiler
- ✓ IGFC (Integrated Gasification Fuel cell Cycle)

1. Introduction

Environment issue

Various heavy metals pollution incidents, such as the pollution of Pb, Cd, Cr, Hg, As Sr, I, Cs and so on, result in serious influences on environment and human health.

Understanding of the emission characteristics to minimize environmental disruption and to improve the efficiency of industrial machinery and plant processes

↓

Importance of real-time and direct monitoring method

Monitoring

- Real Time
- Direct

Laser & Plasma Laboratory

5

1. Introduction

Conventional methods(Industrial analysis, elemental analysis, combustion-gravimetric method…): time consumption(preparation, enrichment and separation); restriction of detection limit

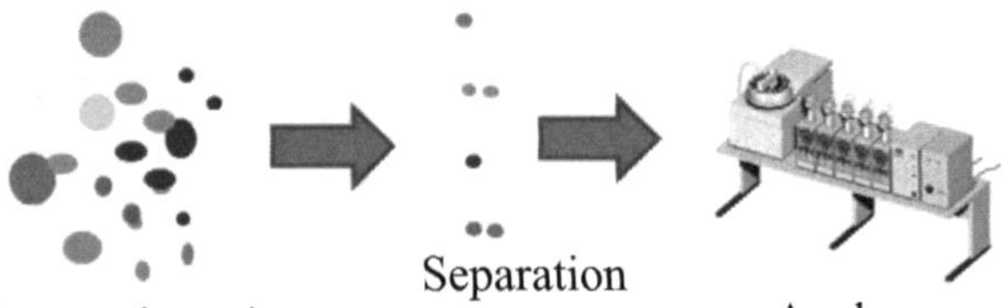

Measured species Separation Concentration Analyzer

Few weeks

This research (LIBS, TOFMS…): rapid detection; direct; high sensitivity

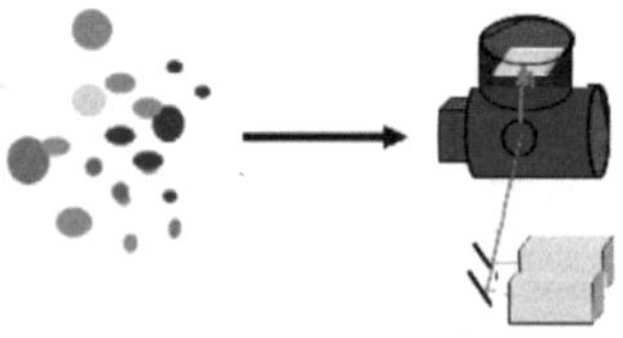

Measured species Analyzer

Few minutes

Laser & Plasma Laboratory

6

2. Theory

qualitative and quantitative analytical detection technique

LIBS---emission

Laser Induced Breakdown Spectroscopy (LIBS):

- fast response, high sensitivity, real-time and non-contact features
- applications: combustion, metallurgy, foods and human, etc.

TOFMS-ion

Time-of-Flight Mass Spectrometry (TOFMS):

- increased sensitivity and rapid analysis
- applications: measurement of hydrocarbons and nanoparticle constituents, waste disposal and treatment plant, medical and pharmaceutical fields

2. Theory

Laser Induced Breakdown Spectroscopy (LIBS)

1) Theory

A laser beam is focused into a small area, producing hot plasma. The material contained in the plasma is atomized, and light is released corresponding to a unique wavelength of each element. The emission intensity from the atomized species provides information on elemental composition.

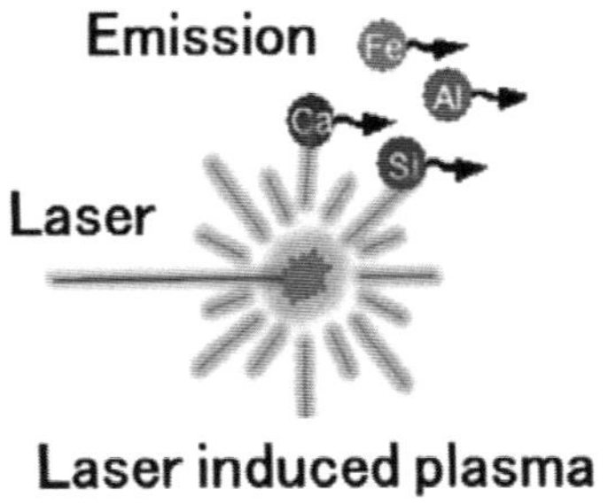

2) Features

- Elemental analysis : %～ppm(ppb)
- Multi-species detection
- High sensitivity(～ppb)

$$I_{(i)} = n_{(i)} \sum_j \left\{ K_{(i),j} g_{(i),j} \exp\left(-\frac{E_{(i),j}}{kT} \right) \right\}$$

2. Theory

Laser Breakdown Time-of-Flight Mass Spectrometry (TOFMS)

1) Theory

Laser breakdown combined with ionization to atomize the molecules and then ionize. A measurement sample is introduced into a vacuum chamber and it is atomized and ionized by laser irradiation. Electric field potential is simultaneously applied for acceleration. The atomic or molecular mass of the sample is proportional to the square of the time of flight.

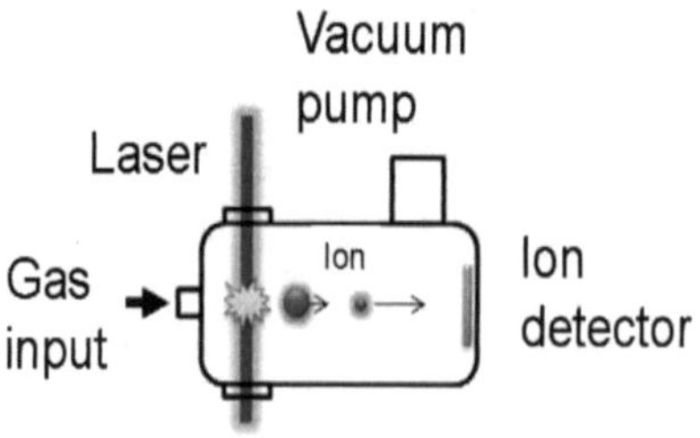

$$m = \frac{2zV}{L^2} t^2$$

2) Features

● High sensitivity (~ppt)

Laser breakdown process

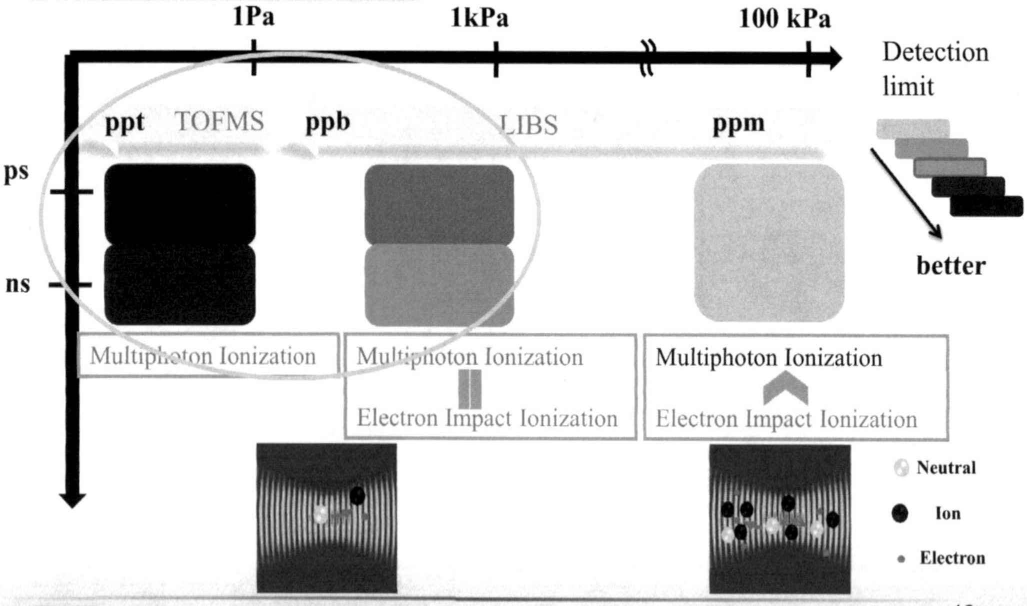

3. Measurement of trace species using low pressure LIBS

Experimental setup

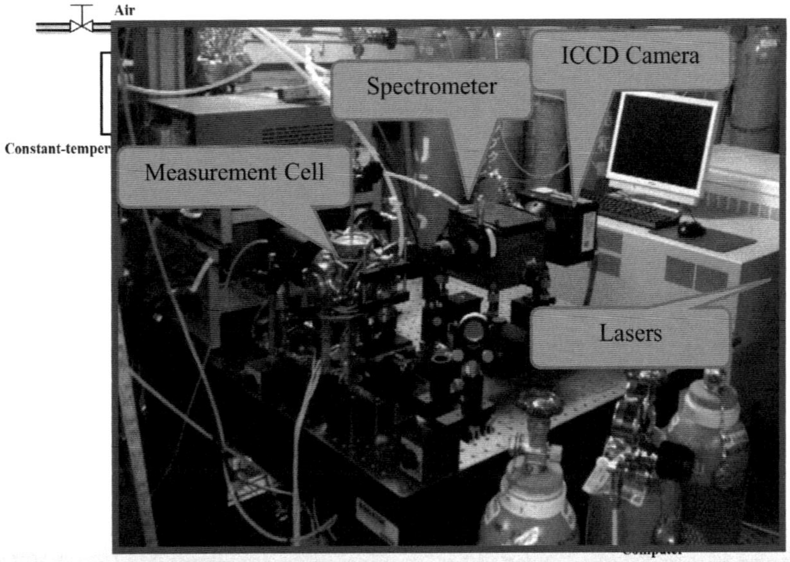

3. Measurement of trace species using low pressure LIBS

Experimental setup

- Experimental parameters: pulse width, laser power, delay time, pressure, concentration

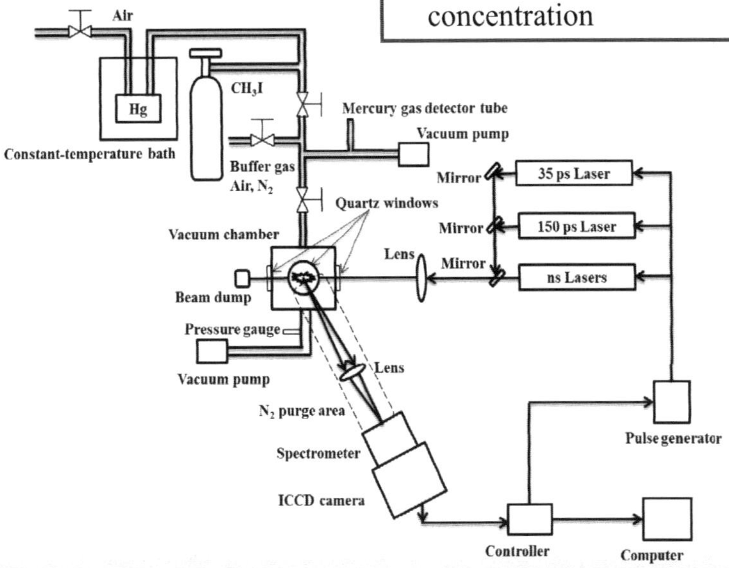

3. Measurement of trace species using low pressure LIBS

Low pressure and short pulse LIBS (Hg)

- Continuum emission
- Coexisting material

nanosecond breakdown

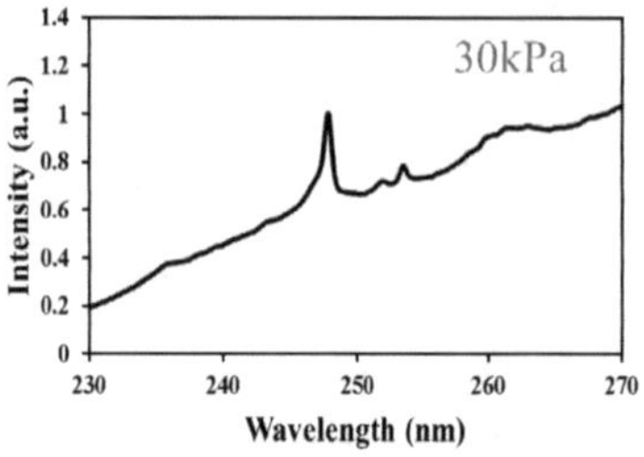

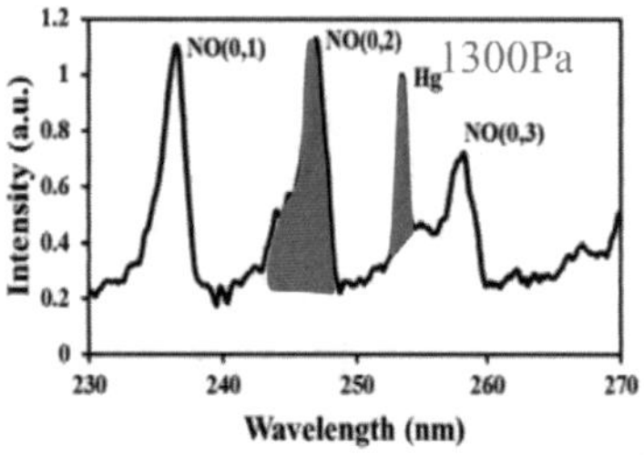

picosecond breakdown

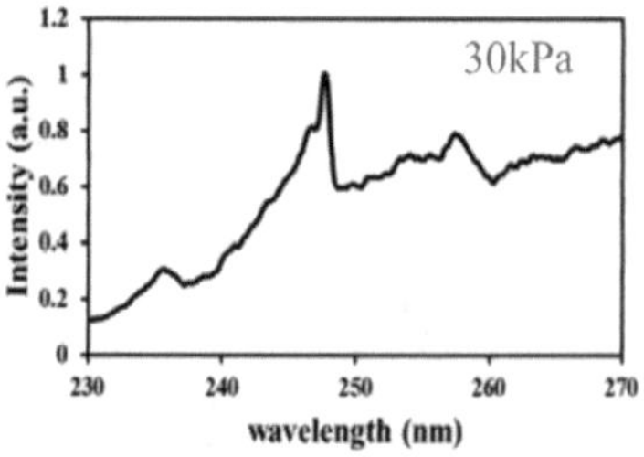

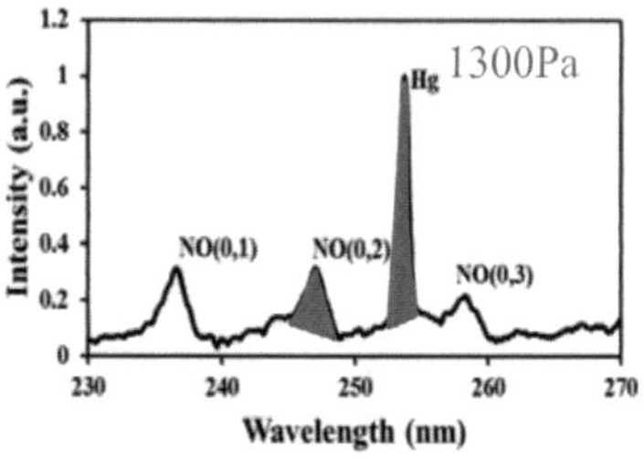

3. Measurement of trace species using low pressure LIBS

Low pressure and short pulse LIBS

- Low pressure
- 532nm breakdown

Pressure dependence

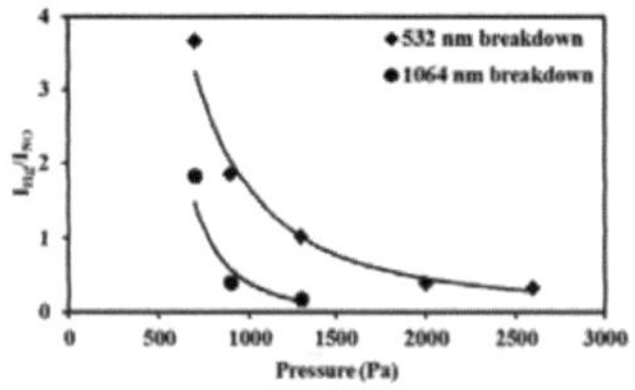

(a) The ratio of I_{Hg}/I_{NO}

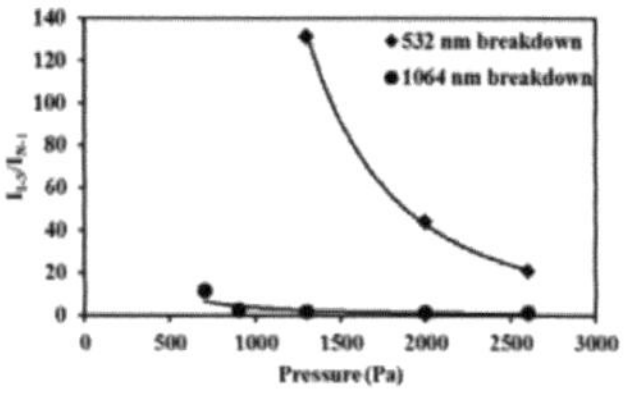

(b) The ratio of I_{I-3}/I_{N-1}

Delay time dependence

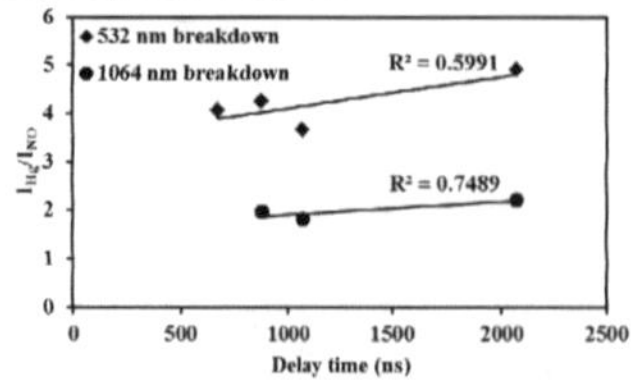

(a) The ratio of I_{Hg}/I_{NO}

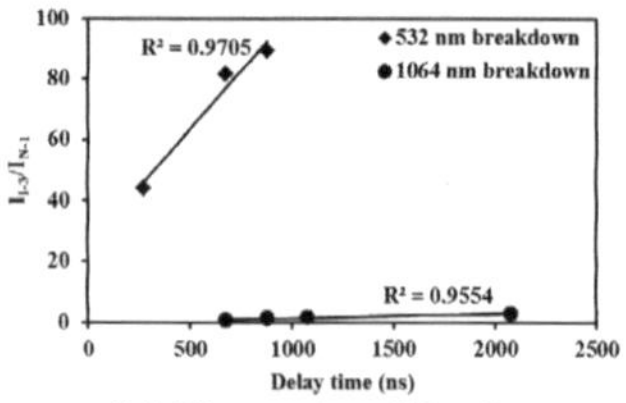

(b) The ratio of I_{I-3}/I_{N-1}

3. Measurement of trace species using low pressure LIBS

Low pressure and short pulse LIBS

Buffer gas effect

- quenching of Hg signal in O_2
- Coexisting atom and molecule emissions

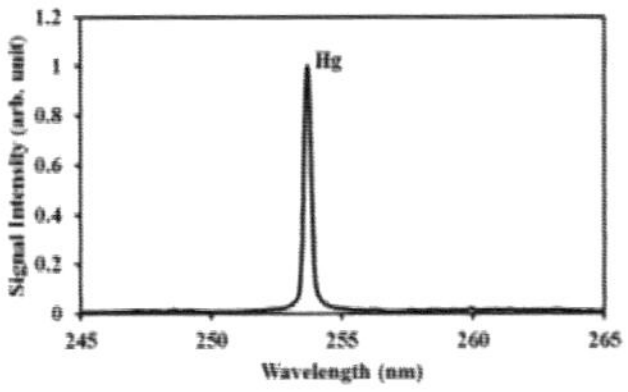

(a) Hg in N_2

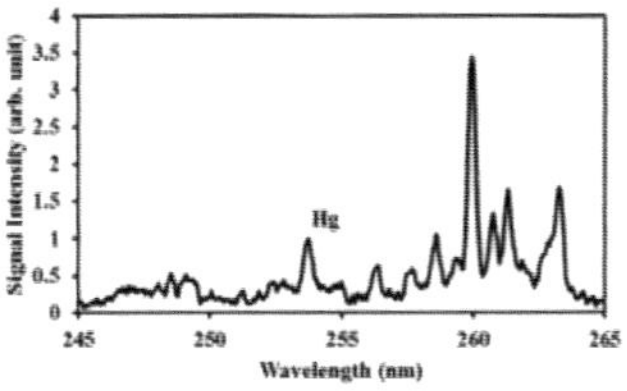

(b) Hg in air

LIBS spectra of Hg in different buffer gases

3. Measurement of trace species using low pressure LIBS

Low pressure and short pulse LIBS

Detection limit of Hg

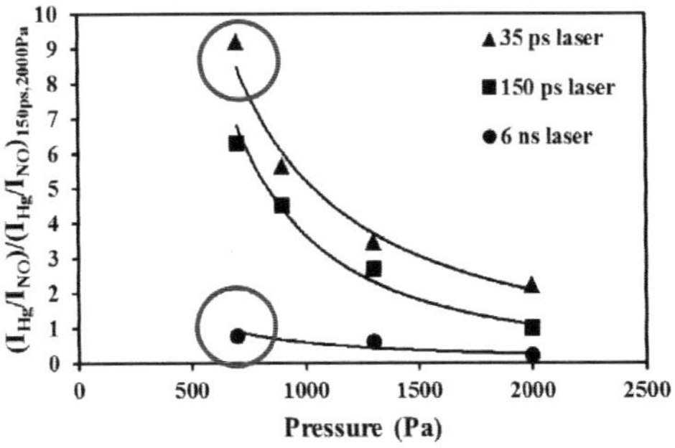

- The increase ratio of I_{Hg}/I_{NO} becomes more than 10 times in the case of picosecond(35ps) laser induced plasma at 700Pa compared to nanosecond(6ns) breakdown.

- Hg in air (35ps): 30ppb(1min).
- Hg in N_2 (ns): 3.5ppb(1min).

I_{Hg}/I_{NO} of ns and ps breakdown at different pressure

3. Measurement of trace species using low pressure LIBS

Low pressure and short pulse LIBS (Sr, Cs)

$Sr[C_5(CH_3)_5]_2$, $Cs(C_{11}H_{19}O_2)$

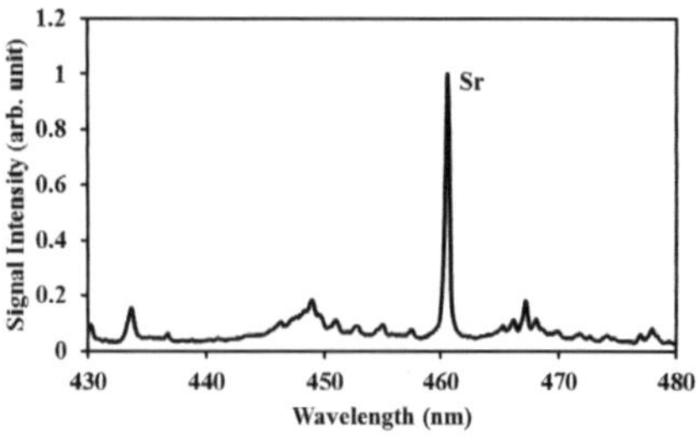

Nanosecond breakdown at 2600Pa

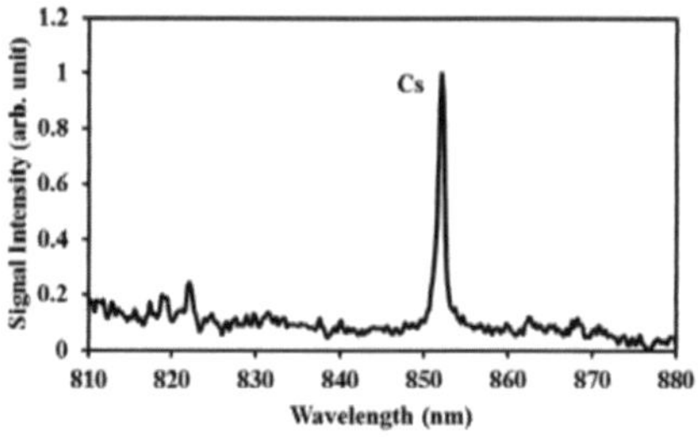

Nanosecond breakdown at 6600Pa

4. Measurement of trace species using LB-TOFMS

Experimental setup

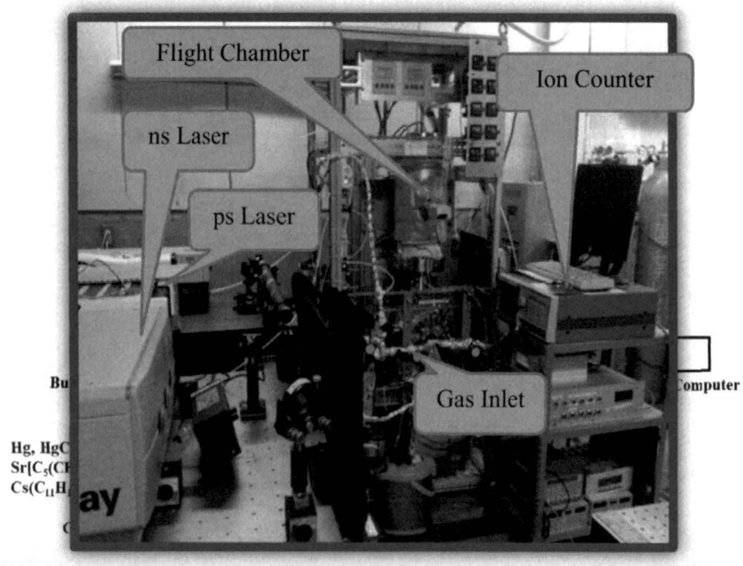

4. Measurement of trace species using LB-TOFMS

Experimental setup

- Experimental parameters: pulse width, laser power, pressure, concentration

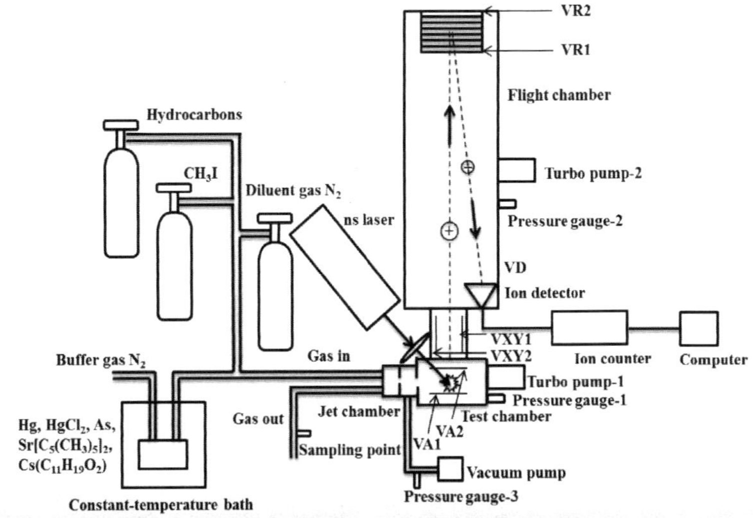

Laser & Plasma Laboratory 19

4. Measurement of trace species using LB-TOFMS

Effects of laser wavelength

- **There is no fragment from hydrocarbons using 532nm and 1064nm breakdowns(nanosecond) around 30-300m/z.**

Target Elements: As, Sr, Cd, I, Cs, Hg, Pb

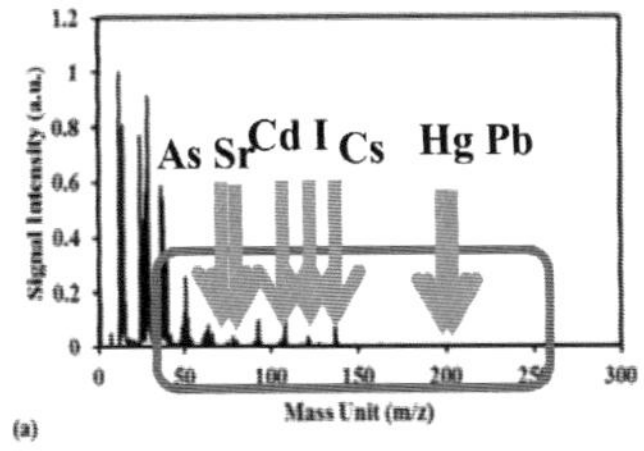

(a)266nm breakdown

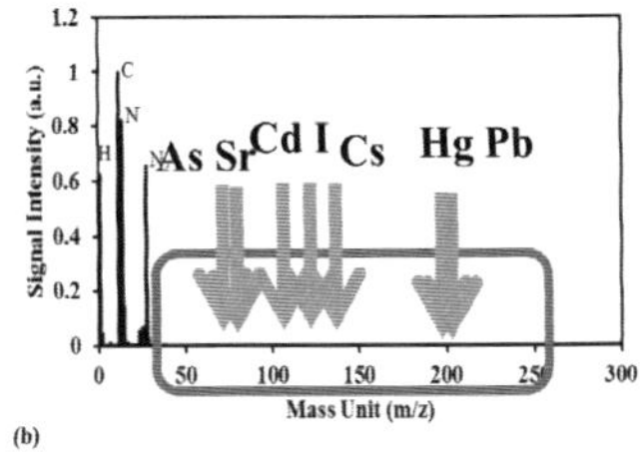

(b)532nm breakdown

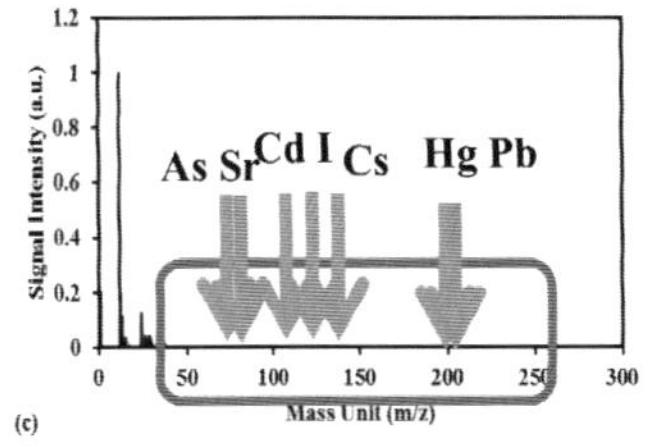

(c)1064nm breakdown

Laser & Plasma Laboratory 20

4. Measurement of trace species using LB-TOFMS

Effects of laser wavelength

Measurement of Hg and CH_3I in N_2(buffer gas)

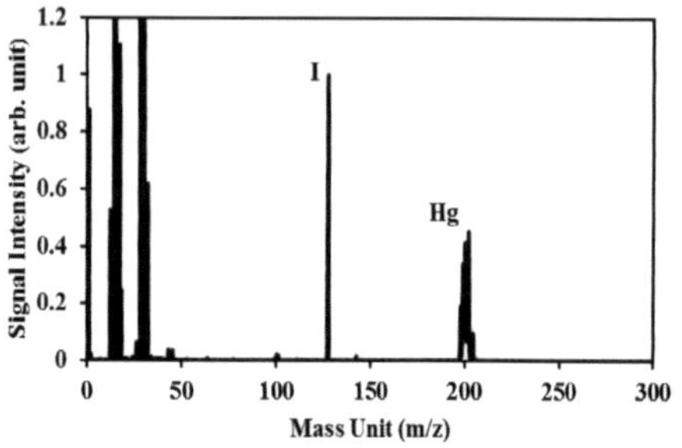

532 nm breakdown of Hg and CH_3I

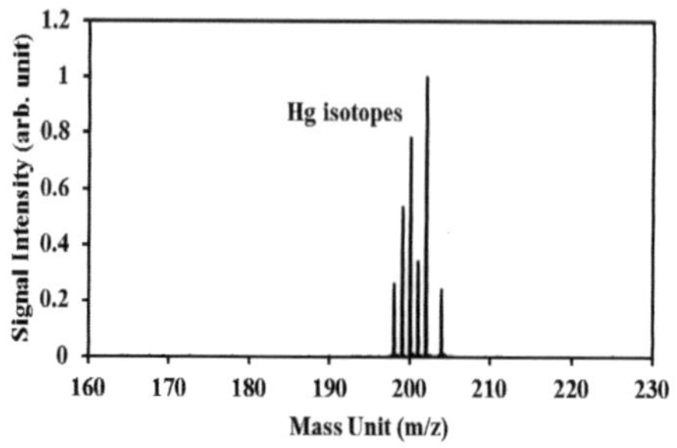

Hg isotopes using 1064 nm breakdown

Laser & Plasma Laboratory

4. Measurement of trace species using LB-TOFMS

Measurement of Hg and CH_3I

Hg concentration is 10 times

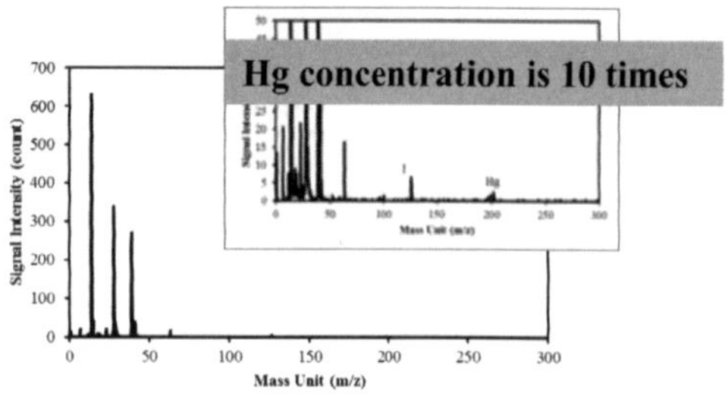

(a) 1064nm breakdown

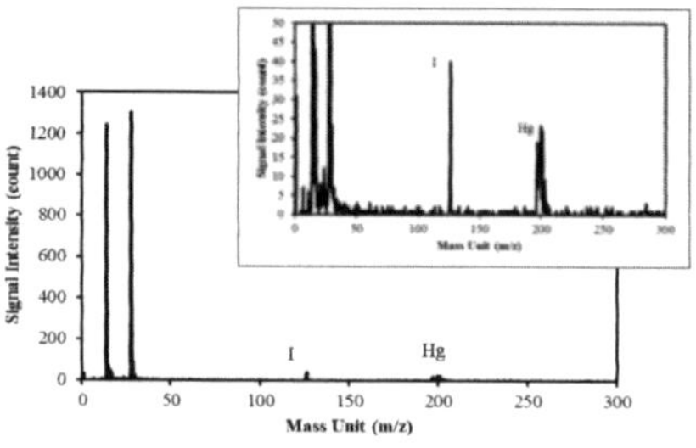

(b) 532nm breakdown

pressure dependence

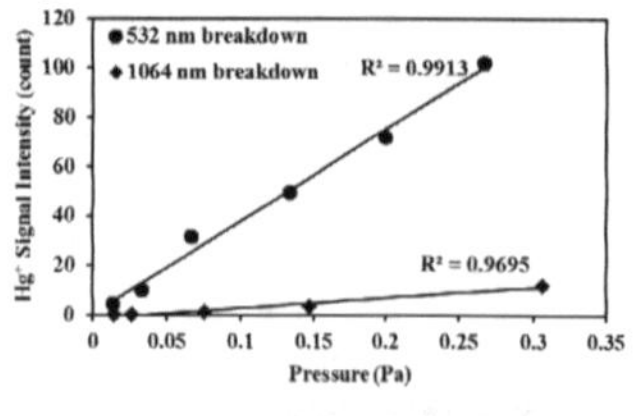

(a) Hg^+ signal

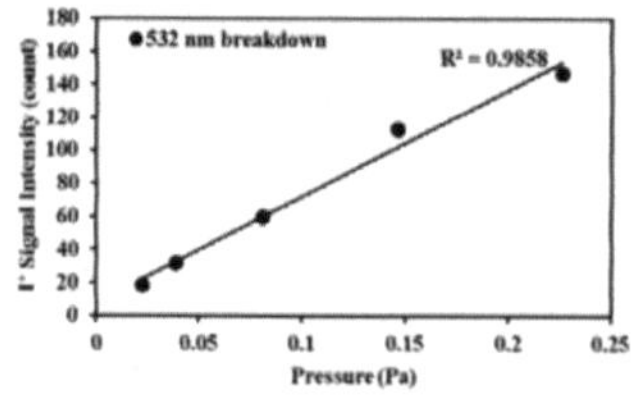

(b) I^+ signal

Laser & Plasma Laboratory

4. Measurement of trace species using LB-TOFMS

Measurement of Hg and CH_3I

laser power dependence

- **Hg and I signal intensity increases and then decreases when increasing the power.**

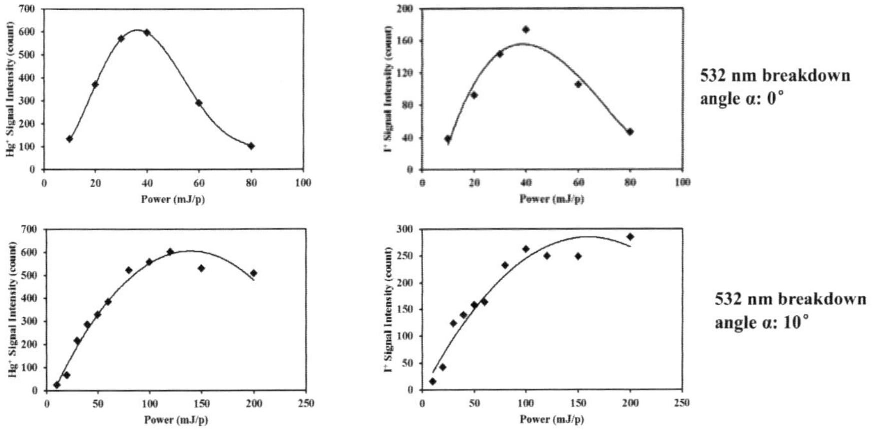

(a) Hg^+ signal (b) I^+ signal

Multi-photon ionization: signal intensity increases as the laser power increases
recombination of Hg ions and electrons under high laser power condition

4. Measurement of trace species using LB-TOFMS

Detection limit

Measurement of Hg and CH_3I in N_2(buffer gas) using *532nm breakdown*

- **The signal intensity of Hg^+ and I^+ can be enhanced 3.8 times and 1.3 times employing 35ps laser breakdown.**

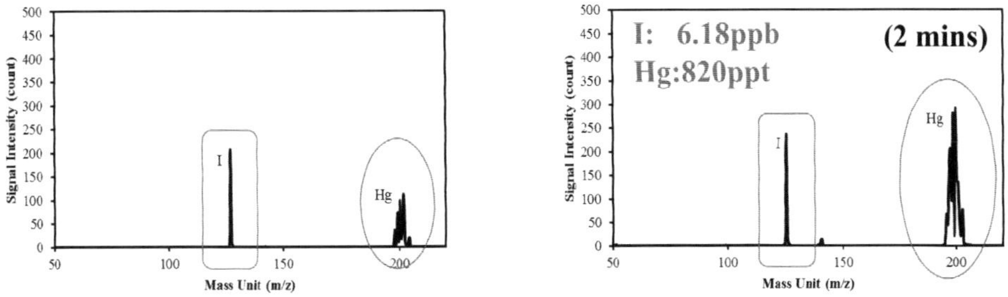

(a) Nanosecond(6ns) breakdown (b) Picosecond(35ps) breakdown

4. Measurement of trace species using LB-TOFMS

Applications of LB-TOFMS

- **Target elements can be detected from complex coexisting materials.**

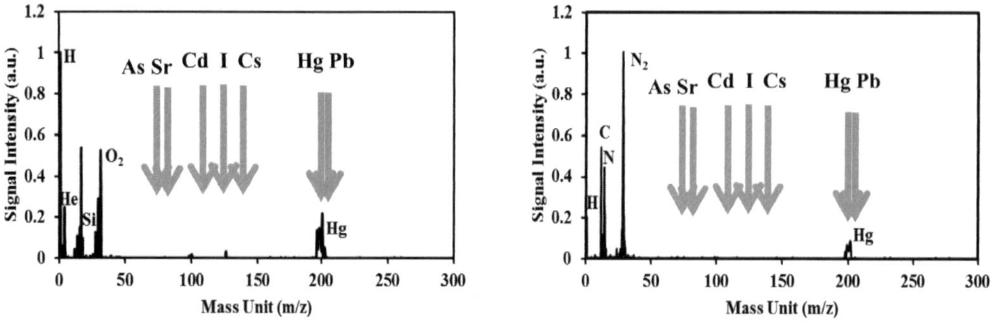

(a) Hg in particles

(b) Hg in hydrocarbons

Fly ash

$p\text{-}C_7H_6Cl_2$, C_7H_8, $C_6H_5C_2H_3$, $p\text{-}C_8H_{10}$, $p\text{-}C_6H_4(C_2H_5)_2$ and $C_6H_3(CH_3)_3$
C_2H_4, C_2H_6, C_3H_8, $n\text{-}C_4H_{10}$ and $p\text{-}C_8H_{10}$

5. Conclusions

- Low pressure and short pulse LIBS was employed to enhance the detection limit of trace species.
  Hg detection limit using picosecond(35ps) breakdown: **30ppb**(in air)

- LB-TOFMS was employed to enhance the detection limit of trace species with few fragmentation from other species.
  Hg detection limit using picosecond(35ps) breakdown: **820ppt**

Future work:

❖ Low pressure LIBS and LB-TOFMS will be applied to the practical applications, such as monitoring systems of power plants and environment according to each measurement requirement.

Thank You
for Your Attention

Based on our previous mechanistic study of the heterogeneous catalytic oxidation of elemental mercury (Hg(0)) vapor over cupric chloride ($CuCl_2$)-based catalyst, the reaction kinetics was studied. The objective of this work is to determine the reaction kinetics for the reaction of Hg(0) with $CuCl_2$. The oxidation reaction has been considered as a gas solid reaction, and a grain model was developed to determine the Hg (0) oxidation reaction under different Hg (0) concentrations and temperatures. A relationship between the local conversion of cupric chloride and reaction time was obtained, and a comparison between simulation and experimental results was made. The detailed simulation and experimental results will be presented.

362286 Investigation on the Chemical Degradation of SOFC Anode Caused By the Trace Level of Sulfur and Other Elements Possibly Contained in Coal Derived Fuel Gas

Abstract:

The integration of gasification and solid oxide fuel cells will be promising direction to meet the demand of clean and efficient utilization of coal for power generation. This study focused on the impact on performance of Ni-YSZ SOFC anode of ppm-level contamination in wet hydrogen or simulated coal syngas (mixtures of H_2 and CO). We observed that the trace level contamination of fuel with H_2S ranged from 0.02 to 2ppm resulted in the irrecoverable performance loss and slight acceleration of chemical degradation of anode which depended on the concentration of H_2S. We also found that the injection of contaminated simulated coal syngas led to the faster degradation compared to the case with hydrogen injection.

DESIGNING ADVANCED ACTIVATED CARBONS FOR ENHANCED MERCURY CAPTURE

Roger Cayton, Jacob Lowring and Joe Wong
ADA Carbon Solutions, Littleton, CO

EPA's Mercury and Air Toxic Standards (MATS) requiring stringent emission controls for coal-fired electric generating unit (EGU) power plants will be enforced beginning April 2015. Mercury control levels will be approaching 85-95% reduction of current non-controlled levels. Activated carbon injection (ACI) using powdered activated carbons (PACs) has been demonstrated commercially as the leading technology for achieving compliance. We will discuss the design of advanced PACs for effective mercury capture and co-benefit integration with the EGU's air quality control system (AQCS). Laboratory and full-scale testing results will be presented for a variety of coal fuel and EGU AQCS processes and operating conditions.

Introduction

Mercury (Hg) is a highly toxic element that occurs naturally, but can accumulate in nature due to emission from human activities, such as the burning of fossil fuels and industrial processes. Roughly 40% of Hg emitted into the environment comes from coal-fired power plants. The U.S. Environmental Protection Agency (EPA) has mandated strict guidelines on the control of hazardous air pollutant emissions from power plants, placing limits on Hg and acid gases, non-Hg metals, dioxins/furans and non-dioxin/furan organics. The Mercury and Air Toxic Standards (MATS) rule requires the use of the maximum achievable control technology (MACT) to reduce power plant emissions of Hg to stipulated low levels, which is equivalent to 80-90+% removal when burning bituminous and Powder River Basin (PRB) coals. The rule applies to both existing and new power plants. MATS compliance must be implemented by April 2015, with the possibility of a one-year extension.

Activated carbon (AC) is widely regarded as the best available control technology for Hg capture. Activated carbon is a porous carbonaceous material, and due to its inherent high pore volume and surface area, activated carbon can capture and sequester a wide variety of impurities in fluid streams, including Hg. To effectively remove Hg from AQCS, activated carbon facilitates three key mechanisms: 1) contact, 2) conversion and 3) capture. Effective contact, conversion and capture are accomplished by tuning the surfaces, pores and particles of the activated carbon.

In this paper, ADA Carbon Solutions' approach to designing advanced PACs will be discussed, starting with the science of Hg capture by activated carbon, and using this fundamental knowledge as a guide to performance optimization leading to enhanced product design. Results from our in-house test to simulate full-scale Hg control that simulates typical ESP mercury capture curves, the Dynamic Mercury Index (DMI), will be shared to show the optimization of three key activated carbon properties: particle size, oxidant additive and pore volume. Finally, a case study will be used to tie laboratory results to full-scale trials. Particle size impact will be discussed using data from a full-scale power plant trial.

In order to have efficient mercury control, the three critical mechanisms of contact, conversion and capture must be accomplished. Due to the design nature of a power plant's emission train, co-benefits from existing AQCS processes, such as a SCR catalyst or a wet flue gas desulfurization scrubber or other, may contribute in full or in part to accomplishing these mechanisms. We will focus our discussion on achieving all these mechanisms using activated carbon.

Discussion

Within the plant's AQCS, PACs may only be in the presence of mercury for a few seconds, and sometimes less than one second. Therefore not only must the three mechanisms take place, but they must happen within the span of a very small time frame. The three mechanisms must also be balanced; there are tradeoffs to each. The challenge is to optimize carbon performance without creating additional balance of plant issues, such as equipment plugging, corrosion, excess halogen in the wastewater, etc.

The first mechanism is contact. Mercury is present in the flue gas in miniscule quantities, generally on the order of parts per billion (ppb) levels. In order to effectively control mercury, activated carbon must first come in contact with or "touch" the mercury species, which is not an easy task when the low mercury concentration and short residence are taken into consideration. The second required mechanism is conversion. Mercury in flue gas streams is generally found in the elemental state (Hg^0 vapor), in oxidized forms (Hg^{1+} or Hg^{2+} species), and as particulates. Elemental mercury is not efficiently adsorbed onto activated carbon. Mercury must be oxidized (Hg^{1+} or Hg^{2+}) in order to enhance its receptivity to activated carbon. The final mechanism that must take place is capture. Once the mercury is contacted and oxidized, it must be securely captured within the activated carbon structure. Activated carbon is able to sequester impurities in fluid streams such as mercury via the process of adsorption, either by physical adsorption (physisorption) through Van der Waals and capillary forces in its pore structure or chemical adsorption (chemisorption) via different surface chemical functional sites. One of activated carbon's advantages over other technologies is its ability to capture mercury and hold onto it tightly, not allowing for reemission back to the flue gas.

Contact, conversion, and capture of mercury are accomplished via tuning the surfaces, pores, and particles of activated carbon. The surfaces of activated carbon include not only the outer surfaces, but also the surfaces found within the internal pore structure of the activated carbon. The surfaces act as a host for the placement of chemical reactants, catalysts and chemical functionalities. The pores of activated carbon are the "holes" that are developed during the activation process. Pores can be compared to the openings found in a piece of sponge. Pores allow the transport of the desired molecules into the carbon, which leads to the capture of molecules with the pore structure. The particles represent the individual units of activated carbon grains. Particles are the transportation medium used to deliver the desired PAC properties to the right place at the right time and subsequent removal of the Hg from the AQCS.

The contact mechanism is most affected by the particles of activated carbon. Particles of activated carbon can be optimized via manipulation of particle size (the median particle diameter on a volumetric basis is generally used, MPD or D50) and particle morphology. Activated carbon is injected into flue gas systems gravimetrically. By increasing the number of particles in a given mass of activated carbon, the statistical probability of "touching" mercury increases dramatically, therefore improving the contact mechanism. The number of particles per mass of activated carbon can be calculated using the MPD, volume, and density of activated carbon. Equation 1 shows the formula to calculate the number of particles per mass of activated carbon, approximated by assuming an ideal spherical particle geometry.

Equation 1. Number of particles per mass of activated carbon calculation

$$\left(\frac{\mathbf{1}\ \boldsymbol{particle}}{\boldsymbol{volume\ of\ particle}\,\frac{\mathbf{4}}{\mathbf{3\Pi r3}}\ (\mu\boldsymbol{m}\mathbf{3})}\right)\left(\frac{\mathbf{1}\ \mu\boldsymbol{m}\mathbf{3}}{\mathbf{1.0}\ \boldsymbol{x}\ \mathbf{10-12}\ \boldsymbol{ml}}\right)\left(\frac{\mathbf{1}}{\boldsymbol{density}\ (\frac{\boldsymbol{g}}{\boldsymbol{ml}})}\right)$$

Table 1 below shows how the nominal number of particles per gram of activated carbon varies when density is held constant at approximately 0.45 g/ml and MPD is changed. Figure 1 graphs the data visually. When the MPD is reduced, the number of particles increases significantly. For example, an activated carbon with a MPD of 12 microns will have approximately 16 times the number of particles per gram as a comparative activated carbon with a MPD of 30 microns. Therefore a carbon with a MPD of 12 microns will have a much greater opportunity to contact mercury. The particle size of activated carbon does have a practical limit, which is dictated by the handling of the material as well as AQCS equipment operability constraints. ADA Carbon Solutions designs PACs with tightly controlled particle size distributions to minimize MPD while maintaining reliable handling and operability. The morphology or shape of activated carbon particles is also important for flow. Balancing particle size and shape leads to an optimum-performing material.

Table 1. Number of particles per gram of activated carbon

MPD (µm)	Estimated Radius (1/2 of the MPD)	Estimated Number of Particles per gram (Billions)
30	15	0.2
25	12.5	0.3
20	10	0.5
18	9	0.7
16	8	1.0
14	7	1.5
12	6	2.5

Figure 1. Number of particles per gram of activated carbon vs. MPD

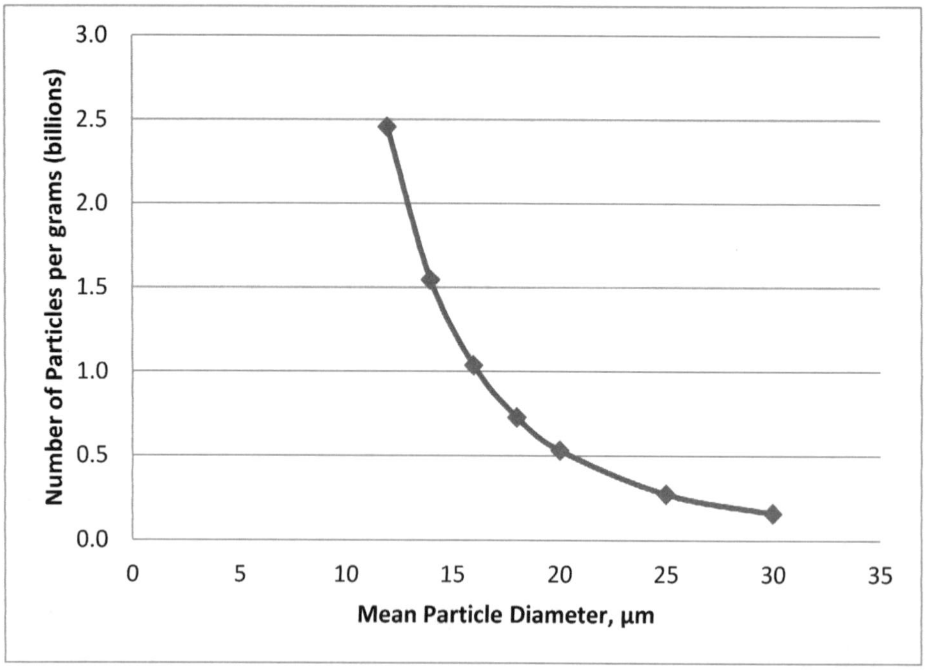

Conversion of mercury is most affected by modifying the surfaces of activated carbon. Chemical reactants and functionalities on the surface of activated carbons can be modified to enhance mercury oxidation. Oxidized forms of mercury behave as Lewis acids leading to strong chemisorption to Lewis base sites on the activated carbon surface. Elemental mercury is unable to interact with the activated carbon in the same way as oxidized mercury. Halogens are effective oxidants for mercury. Chlorine, bromine and iodine are all used as oxidants; however, bromine is most often used due to its high effectiveness and lower cost. Historic full-scale power plant data have shown that oxidation is necessary for effective mercury control. Nelson, et. al. showed that the performance of PAC with an oxidant additive or treatment is on order of 2-3 times more effective in Hg capture than non-brominated PAC[1]. This trend can vary depending on plant configuration, as well as coal composition.

Figure 2. Pore Structure of Activated Carbon

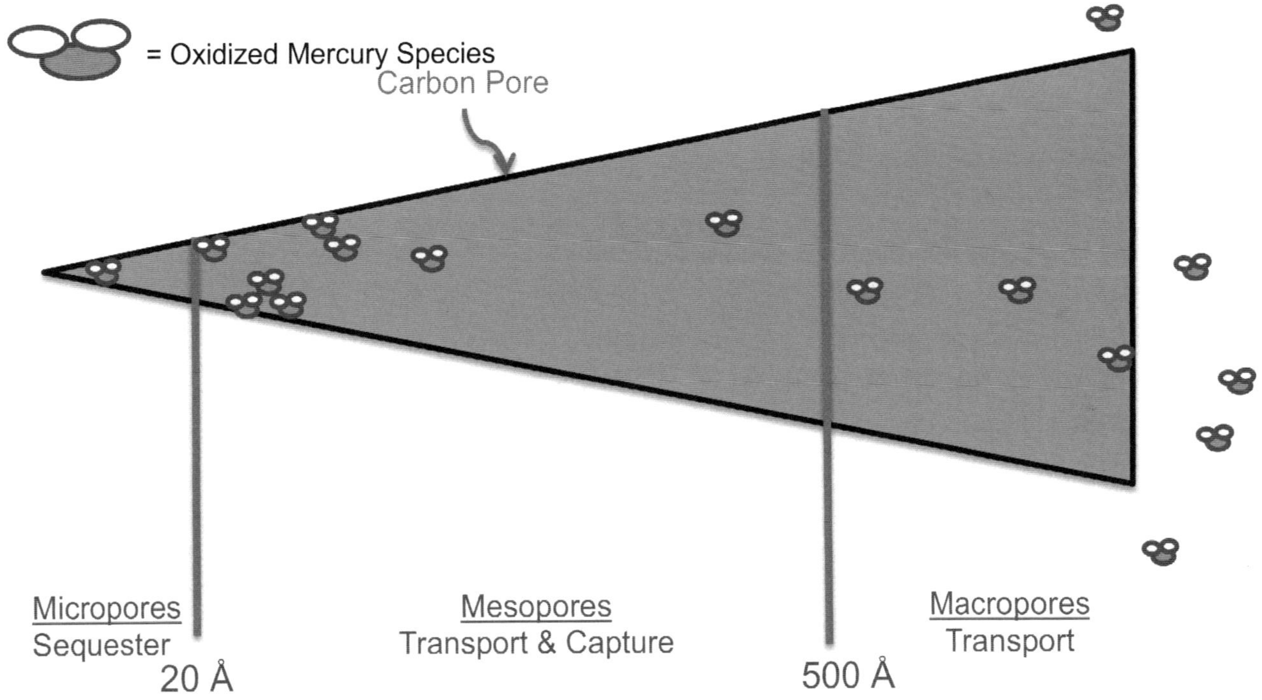

Secure capture of mercury is most affected by the pores of activated carbon. Activated carbon has an inherently large amount of pore volume (greater than 0.2 cc/g) allowing for diffusion of mercury into the pore structure. Pore volume is necessary; however, the distribution of pore sizes within activated carbon is equally important to mercury capture. Pore size is broken down into three categories as designated by IUPAC: Micropores 0-20 angstroms in size; Mesopores 20-500 angstroms in size; and Macropores greater than 500 angstroms in size. A distribution of macropores, mesopores, and micropores is necessary for effective mercury control. This concept can be visualized using the butterfly net analogy (Figure 2).

Macropores, like the opening of a butterfly net, allow initial contact and entrance of mercury into the activated carbon pore structure. Mesopores provide opportunities for initial mercury capture and transport further into the carbon pore structure. Finally, micropores sequester the mercury, preventing reemission. The right amount of total pore volume and the appropriate pore size distribution provide an optimum activated carbon for mercury control.

One of the in-house methods that ADA Carbon Solutions has developed to rapidly optimize powdered activated carbons is the Dynamic Mercury Index (DMI)[2]. Traditionally the capability of activated carbons to adsorb mercury has been modeled using fixed bed testing. In fixed bed testing, a bed of activated carbon is arranged whereby a simulated flue gas is passed through an activated carbon bed and the amount of mercury retained on the sorbent is measured. As such, the fixed bed test is biased towards the total capacity of the activated carbon. Since the carbon is stationary, a long contact time is simulated, allowing maximum adsorption of available mercury. However, with activated carbon injection into power plants, mercury adsorption is kinetically controlled as the contact is in the range of two seconds or less. Therefore, activated carbons never reach a capacity limitation and must be designed for maximum surface site oxidation/chemisorption of various mercury species for quick and efficient capture. ADA Carbon Solutions sought to develop a test method that would simulate the actual kinetic environment where mercury capture takes place in a flue gas environment. The best way to develop effective powdered activated carbons for flue gas mercury control was found to be a dynamic pilot plant (the DMI, See Figure 3. for representative schematic) where contact times and gas conditions simulate that of an EGU emissions train.

The DMI simulates the dynamic conditions found in a flue gas system and provides removal efficiency and speciation to quantify conversion and capture. Because it is a dynamic test, it also reflects the benefits of smaller particle size and modified shape for improved contact. The particle size benefit is difficult to quantify in a fixed bed test since total capacity may be similar to a larger size particle. The short residence time DMI test also reflects the relative contributions of surface area and pore volume compared to surface active sites. Via the DMI test methodology, a higher DMI number translates to a better performing material for mercury control.

Figure 3. Dynamic Mercury Index Instrument

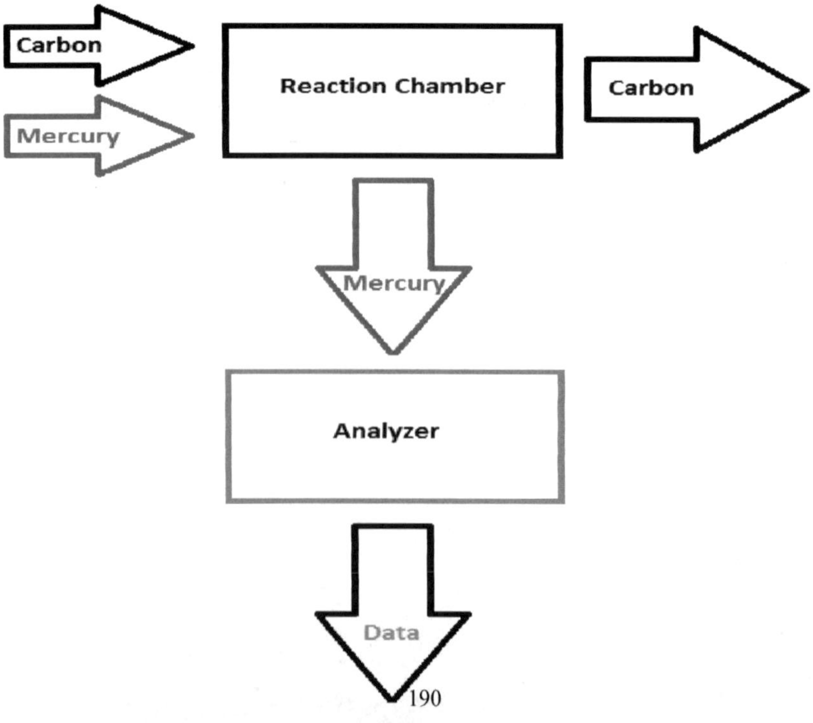

The DMI has been used to define the impact of many carbon properties, including particle size, oxidation effects and pore size distribution. Figure 4 shows the results of lab-scale testing using the DMI to test the effect of MPD on activated carbon performance. PACs tested contained a constant level of mercury oxidant to monitor the effect of only changing particle size. The graph shows that increasing the MPD of the powdered activated carbon leads to a lower DMI value, and therefore lower Hg capture performance. Conversely, decreasing mean particle diameter gives a higher performing material. Note that the general shape of the graph reflects that of Figure 1 which shows that decreasing particle size dramatically increases the number of particles present leading to greater opportunity of mercury content. The particle size optimization graph has a generally lower slope then the theoretical curve in Figure 1 likely since the calculation assumes uniform particle size throughout the material, when in reality the material contains a distribution of varying particle sizes and shapes.

Figure 5 shows the results of lab-scale testing using the DMI to test the effect of mercury oxidant additive on activated carbon performance. Carbons of comparable particle size were tested both with and without oxidant. As discussed previously, un-oxidized mercury is different to capture. The DMI test shows the carbon sample with oxidant dramatically outperforming the carbon sample without oxidant.

Figure 4. Particle Size Optimization via DMI

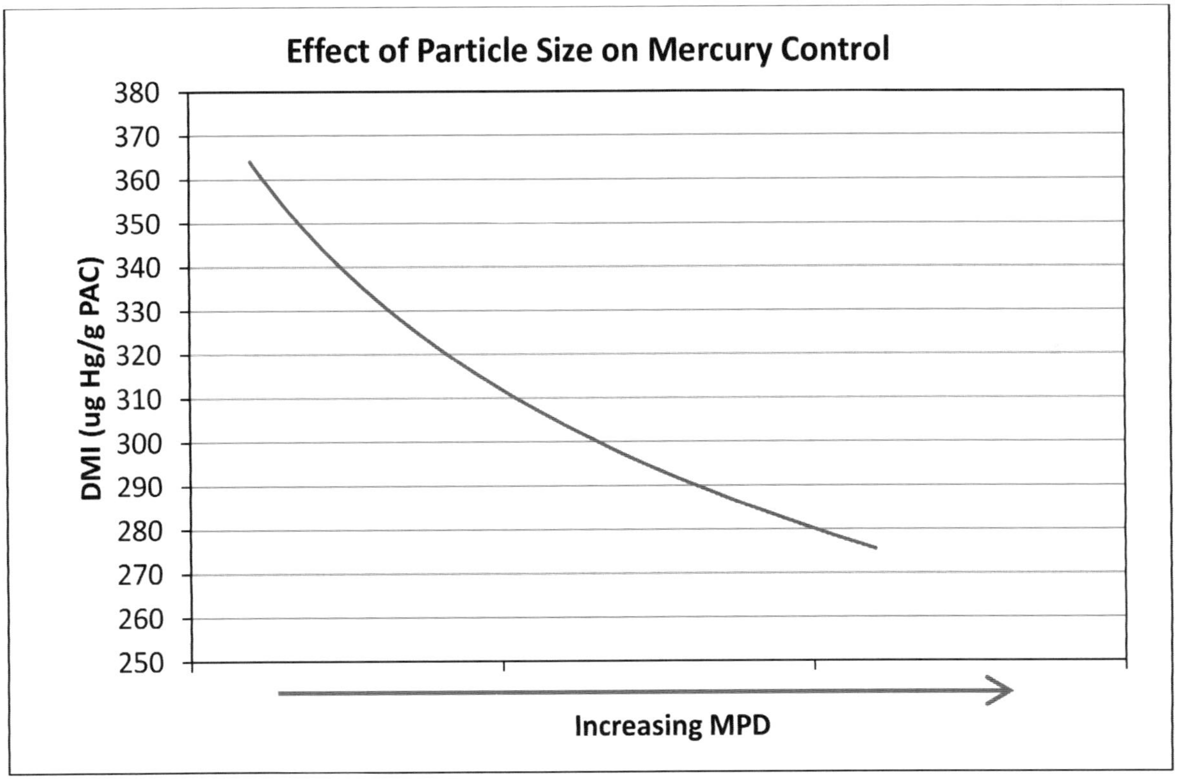

Figure 5. Oxidant Optimization via DMI

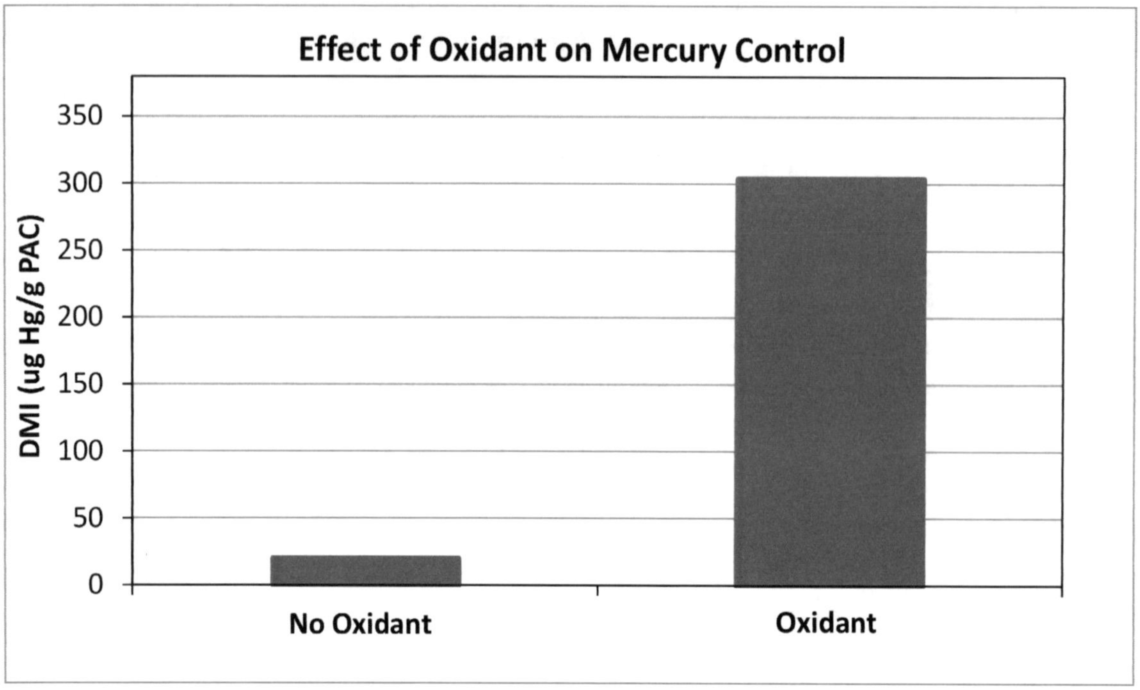

Figure 6 shows the results of lab-scale testing using the DMI to test the effect of pore volume on activated carbon performance. Carbons were compared based on their total pore volumes. Each carbon was comparable in particle size and oxidant content. In Figure 6, Carbon 1 contained the larger amount of total pore volume and outperformed Carbon 2, which contained a smaller amount of total pore volume. These results indicate that total pore volume is important to mercury capture. However as stated previously, the distribution of pore sizes is also vital. By designing a powdered activated carbon with the correct levels of total pore volume and the appropriate pore size distribution, a carbon can be tuned for ideal Hg capture performance.

Figure 6. Total Pore Volume Optimization via DMI

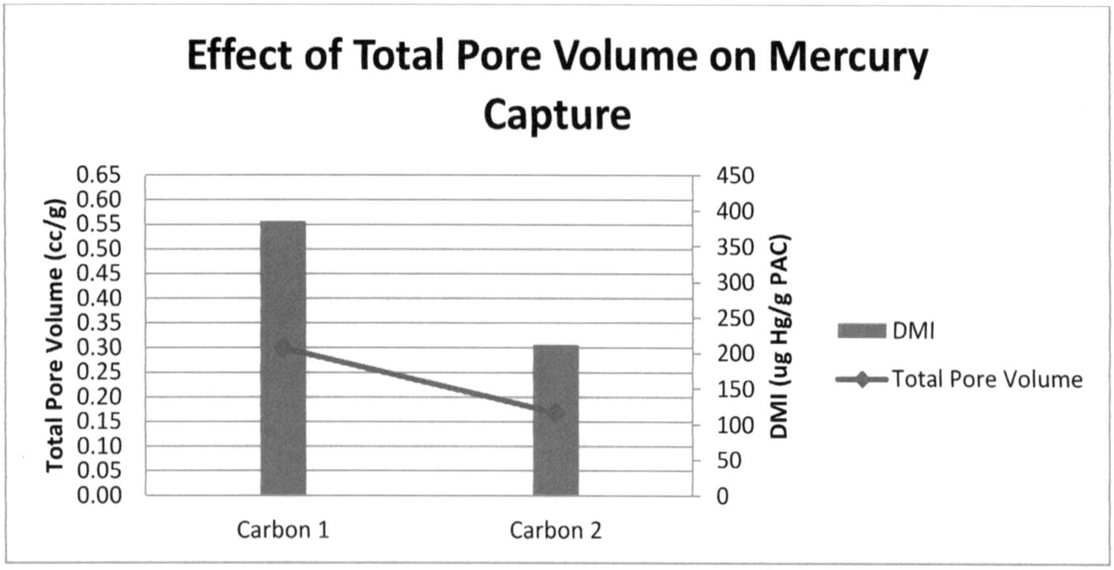

ADA Carbon Solutions has conducted testing of activated carbon injection at numerous full-scale power plants over the past several years. This testing of novel products has been used to confirm lab-scale PAC performance results as well as to validate the ability of the DMI to predict behavior in large-scale applications. The final portion of this paper will be used to review a full-scale case study showing the impact of PAC particle size. The test was conducted on a roughly 200 Megawatt unit, burning Powder River Basin (PRB) coal. The emission train contained an air preheater and cold side Electrostatic Precipitator (ESP) leading to the emission stack. Powdered brominated activated carbon was injected upstream of the ESP. Mercury removal was observed using a mercury continuous emissions monitoring system (CEMS). In order to achieve compliance the unit needed to meet the United States MATS Compliance Rate of 1.2 lbs of mercury per trillion BTUs of energy out of the stack (lb Hg/TBtu). Figure 7 shows the results of injecting the two brominated powdered activated carbons with different particle sizes into the unit. Carbon 1 is approximately 25 microns in MPD and Carbon 2 is approximately 12 microns in MPD. As the figure shows, Carbon 1 met compliance at roughly 165 lbs of activated carbon injected per hour. Carbon 2 met compliance at roughly 75 lbs of activated carbon injected per hour, a 55% reduction in activated carbon injection rate. These results corroborate with the theory and lab data presented earlier showing that decreasing particle size results in an increased number of particles and therefore improved probability of mercury contact.

Figure 7. Power Plant Test – Effect of Particle Size on Mercury Control

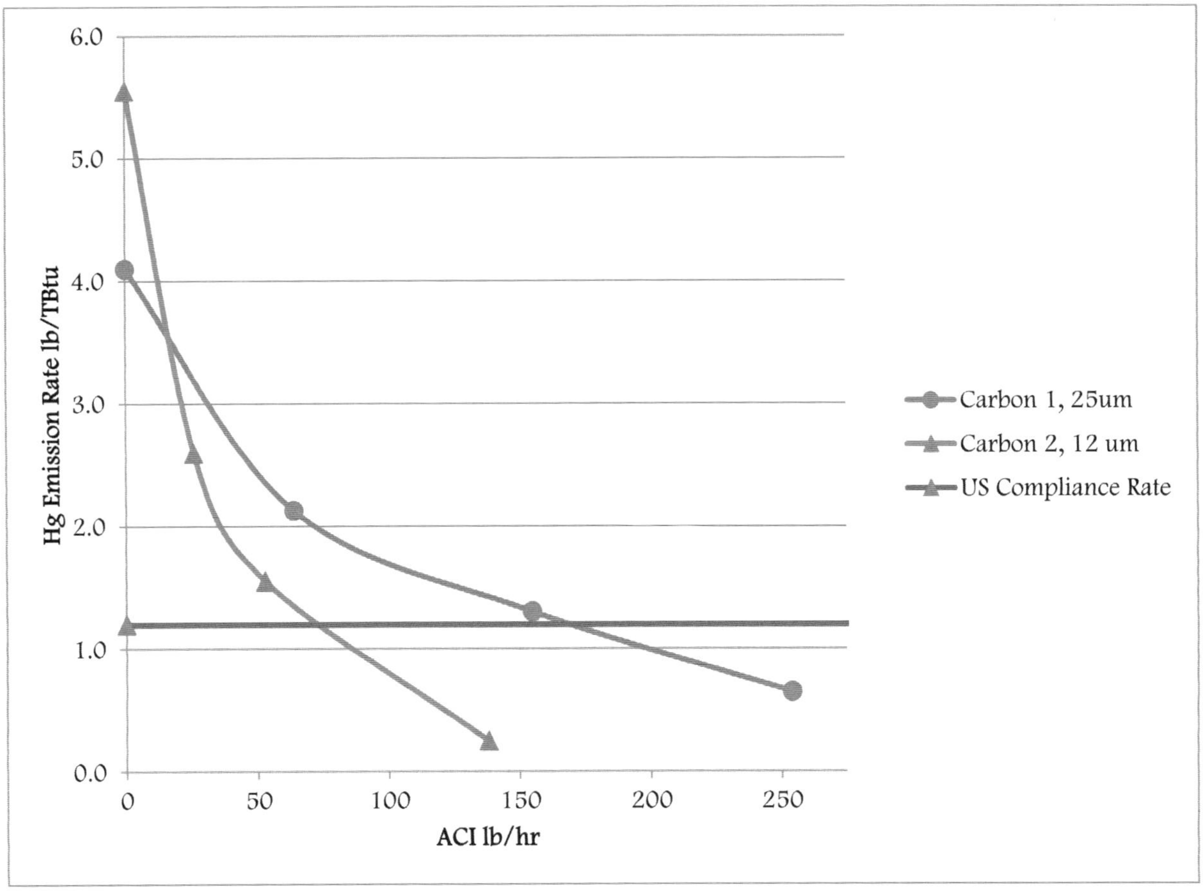

Summary

This paper has presented ADA Carbon Solutions' approach to designing enhanced PACs using scientific rigor. The fundamentals of mercury control using activated carbon to enhance contact, conversion and capture were described, as well as how activated carbon surfaces, pores and particles can be tuned to improve mercury capture. Lab data was presented showing how ADA Carbon Solutions' lab DMI test, which simulates dynamic mercury control in power plants, can be used to optimize activated carbon properties, including particle size, oxidant additive and pore volume. Lab results were further validated against the theory of mercury capture via a full-scale power plant trial, reinforcing a rigorous technical path to enhancing PAC performance.

1. Nelson, S., R. Landreth, Q. Zhou, & J. Miller, 4th DOE-U.S. EPA-AWMA Combined Power Plant Air Pollutant Control "Mega" Symposium, Washington, D.C., Aug. 30-Sept. 2, 2004.
2. Nicholas Lentz, Chris Vizcaino, Sheila Glesmann and Joe Wong, "Advancing PAC Innovation & Quality ADA Carbon Solutions Dynamic Mercury Test," 2014 EUEC, Phoenix, AZ, February 5, 2014.

361838 Sustainable Materials for Mercury Control

Abstract:

USA Environmental Protection Agency recently signed a Clean Air Act rule that will reduce the allowable mercury emissions for electric power plants. Here, we present results obtained from laboratory studies of mercury uptake from a simulated flue gas using unactivated biochars made from agriculture and biorefinery residues. The residues were pyrolyzed at four different temperatures (350, 500, 650, and 800°C). The results show that biochars made from almond shells and cotton seed hulls had very little affinity for mercury. The performance of lignin-based biochar showed clear dependence on pyrolysis temperature, with the biochars made at either 500 or 650°C performing the best. The same was true for biochars created from chicken manure (containing lignocellulosics), and biochars created at higher temperatures out-performed those made at lower temperatures. Overall, the best performance was seen with chicken-manure-based biochar created at 800°C. This material removed 80-98% of the mercury from the simulated flue gas.

Steam activation of the manure-based biochar improved the performance and 100% of the mercury was routinely removed; however, the source of the manure was found to be important. Manure from turkeys did not have the same affinity as manure from chickens. The presence of chicken house bedding material in the manure had an effect on surface area properties but not on the mercury uptake.

Cadmium, Phosphorus and Antimony Adsorption on Copper-Palladium Alloy Films

Karen Uffalussy, James Miller, Bret Howard, Dennis Stanko and Evan Granite

Abstract

The adsorption of phosphine, cadmium and antimony on copper-palladium alloy films at 288°C was studied by scanning electron microscopy with energy-dispersive X-ray methods (SEM-EDX) characterization. The composition spread alloy film (CSAF) coupons were exposed to cadmium, phosphine or antimony in nitrogen in a small flow reactor. Trends for cadmium, phosphorus and antimony uptake as a function of alloy composition are being determined. The results have potential consequences for the design of sorbents for the capture of phosphorus from fluid streams, and for the stability of palladium-copper alloy membranes employed for hydrogen separation from coal-derived syngas.

Uncertainty Quantification of Properties Models: Application to a CO_2-Capture System

Josh Morgan[a], Benjamin Omell[a], Debangsu Bhattacharyya[a], Charles Tong[b], David C. Miller[c]

[a] *Department of Chemical Engineering, West Virginia University, Morgantown, WV 26506, USA*

[b] *Lawrence Livermore National Laboratory, Livermore, CA 94550, USA*

[c] *National Energy Technology Laboratory, 626 Cochrans Mill Rd, Pittsburgh, PA 15236, USA*

Uncertainties in properties models can significantly affect the results obtained from process simulations. If these uncertainties are not quantified, optimal plant designs based on such models can be misleading. With this incentive, a systematic, generalized uncertainty quantification (UQ) methodology for properties models is developed in this work. Starting with prior beliefs about parametric uncertainties, a Bayesian method is used to derive informed posteriors by using the experimental data. To reduce the computational expense, surrogate response surface models are developed. For down-selecting the parameter space, an intuitive approach based on a sensitivity matrix has been developed. As significant uncertainties exist in the properties models of the electrolyte systems, the methodology is deployed to the properties models of an MEA-CO_2-H_2O system.

In particular, UQ of the models for density, viscosity, and surface tension for the MEA-H_2O-CO_2 system is analyzed. All three property models use correlations that predict the mixture property as a function of temperature, CO_2 loading, and amine concentration. The existing density model in the open literature is altered to account for electrolyte speciation. An existing surface tension model is altered to represent the mixture surface tension as a continuous function of all three independent variables and to improve the model accuracy over a wider operating region. The deterministic models for all three properties are developed by regressing to the experimental data for ranges of temperature (298.15-353.15 K), CO_2 loading (0-0.56 mol CO_2/mol MEA), and amine weight fraction on a CO_2-free basis (0.2-0.4), although these ranges vary slightly for different physical properties. The UQ analysis is found to provide interesting information about uncertainties in the parameter space. The sensitivity matrix approach is also found to be a valuable tool for reducing computational expense.

Characterization of Dialkylimidazolium Bistriflamide Ionic Liquids for CO_2 Capture

There has been considerable interest in examining the potential of room temperature ionic liquids for carbon capture processes. In this study, the solubility and diffusivity of CO_2 in 12 different 1,3-difunctionalized imidazolium bistriflamide ionic liquids (ILs) were determined using gas uptake measurements. The density, viscosity, conductivity and self diffusivity of the ILs were also measured. The ionicity of the ILs were quantified in terms of the Haven ratio, calculated using the self-diffusion coefficient and conductivity. Relationships between CO_2 solubility and transport in these ILs and the IL physical properties have been developed. The Henry's Law constant for CO_2 in the ILs showed a strong correlation to the square of the difference in the solubility parameters of the ILs and CO_2, consistent with regular solution theory. A stronger correlation was observed when both the IL solubility parameter and the Haven ratio were included in the correlation. CO_2 diffusivities in the ILs were found to be linearly related to the self-diffusion coefficient of the IL cations. Interestingly, structural isomers of ILs were found to have the same CO_2 solubility (when expressed as CO_2 mole fraction), but distinct CO_2 diffusivities. The product of solubility and diffusivity (i.e., permeability) was found to correlate well with IL viscosity alone, despite the fact that solubility and diffusivity were each found to depend upon both IL viscosity and molar volume.

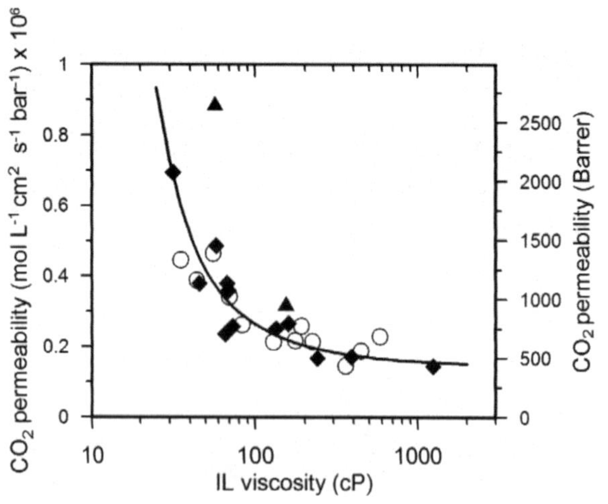

CO_2 permeability was observed to correlate well solely with ionic liquid viscosity

390908 Long Term Durability of Dolomite-Derived MgO-Based Sorbent for CO2 Capture in IGCC

Abstract:

New generation of coal power plants such as Integrated Gasification Combined Cycle (IGCC) reduces the emissions of CO2

efficiency losses, it is desirable to remove CO2 in the temperature range of 300° to 450°C, which makes regenerable MgO-based sorbents ideal for such operations. The sorbents have high reactivity and good capacity toward CO2 absorption in the temperature range of 300–450 C at 20 atm. and can be easily regenerated at the temperature range of 500–550 C. The effect of cycling on CO2 capture capacity of MgO-based sorbent prepared from natural dolomite was experimentally investigated in this work. The rates of carbonation and regeneration reactions were studied. Series of carbonation/regeneration cycles (up to 24) have been carried out in a dispersed bed reactor to determine the effect of various variables on long term durability of the sorbent. The gradual loss of CO2 sorption capacity appears to be mainly due to loss potassium (a carbonation reaction promoter) in the cyclic process. Durability of the sorbents improves in the presence of steam, which is likely due to the favorable changes in the pore structure of the sorbents. A variable diffusivity shrinking core model has been used to fit the experimental data of the sorbent carbonation and regeneration in the cyclic process with excellent accuracy.

per unit of power generated. In IGCC processes, to avoid

Dynamic Modeling and Control of an Integrated Solid Sorbent Based CO_2 - Capture Process

Benjamin Omell[a], Debangsu Bhattacharyya[a], Stephen E. Zitney[b], David C. Miller[c]

[a] *Department of Chemical Engineering, West Virginia University, Morgantown, WV 26506, USA*

[b] *National Energy Technology Laboratory, Morgantown, WV 26507, USA*

[c] *National Energy Technology Laboratory, 626 Cochrans Mill Rd, Pittsburgh, PA 15236, USA*

Post-combustion CO_2 capture is a promising method to control the large greenhouse gas emissions emitted by existing large coal-fired power plants. Solid-Sorbents are a promising alternative to MEA solvent systems due to high CO_2 adsorption and low regeneration cost. This presentation will focus on a solid-sorbent CO_2 capture system that consists of a two-stage bubbling fluidized bed (BFB) adsorber-reactor for CO_2 capture, a moving bed (MB) regenerator, a CO_2 compression train, and balance of the plant. These models are being developed under the auspices of the U.S. Department of Energy's *Carbon Capture Simulation Initiative* (CCSI) that is focused on accelerating the development and commercialization of post-combustion CO_2 capture technologies.

The BFBs are well accepted by industry due to their higher mass and heat transfer rates. The one-dimensional, two-phase model of the BFB is flexible and can be applied over a wide range of operating conditions for gas and solids flowrates and compositions, as well as operating temperatures. Mass and energy conservation equations are written for all the chemical components in both the phases. For modeling purposes, each BFB is divided into three regions namely bubble, cloud-wake, and emulsion. Due to the physics of these three regions, mass and heat exchange has been modeled in the gas phase alone between the bubble and cloud-wake regions whereas mass and heat exchange has been considered in both the gas and solid phases between emulsion and cloud-wake regions.

Moving beds (MBs) can provide a true countercurrent contacting pattern for gas and solid systems. The MB model is a rigorous, non-isothermal, 1-d, two-phase model that was developed in both gPROMS and Aspen Custom Modeler (ACM) and is flexible enough to be used to model both adsorption and regeneration within a wide range of feasible operating conditions. The reaction is exothermic, so an integrated heat-exchanger is used in order to cool the MB adsorption column, and likewise in order to heat the regenerator. In the MB model, the gas flow has been modeled by considering convective flow with axial dispersion. The hydrodynamics of MBs have not been widely studied, so most coefficients have been derived by analogy with correlations for fixed and fluidized bed systems.

Since coal-plants produce large flows of flue gas and MBs require a relatively small gas velocity, overly large and cost prohibitive amount of MBs would need to be considered. Hence a series of regenerator MBs are used with a CO_2 draw-off between stages, thereby decreasing the superficial velocity. As this is a dynamic pressure-driven system, the CO_2 draw-offs present a challenging control problem as the CO_2 draw-offs are actually moving boundary conditions. To this end, effective control strategies are developed to operate the system efficiently in the face of typical disturbances.

In addition, the dynamic model of an integral gear centrifugal CO_2 compression system has been developed and coupled with the solid-sorbent CO_2 capture system. The compressor train compresses the CO_2-rich gas from about the atmospheric pressure to the supercritical pressure for pipeline transport. The process dynamics are studied by simulating a number of disturbances such as inlet flow, temperature, and composition of the flue gas.

390699 Numerical Simulation of a Pyrolysis and Re-Burning Combinative Low-NOx Grate Boiler

Abstract:

With investment in the coal industry, China retains its current position as the leading global producer and consumer of coal, even as it endeavors to diversify its energy structure. The consumption of coal in china is largely in power production, aside from this, there is a lot of industrial use along with a comparatively small amount of domestic use. However, because of the limitation of present economy conditions in some underdeveloped areas, domestic coal combustion for heat supply and cooking still relies on small grate boiler, of which efficiency is comparatively low and has a high pollutant emission rate. When coal is burned, NO_x mainly includes nitrogen oxide (NO) and nitrogen dioxide (NO_2) will be produced. These pollutants initiate reactions which result in the production of photochemical smog, ozone and acid rain. That is why NO_x has drawn much attention of all over the world.

According to these low-NO_x combustion techniques which are widely used in industry and domestic area, this paper presents an advanced grate combustion system, which combine pyrolysis and re-burning low-NO_x techniques. This technique refers to a new design of a grate boiler, which works with three different parts. Firstly, the pyrolysis chamber gives sufficient residence time for reduction gas generation and low NO_x emission, while the main combustion chamber as the second parts can acts as a well-stirred complete combustion zone. Meanwhile, introduce industrial exhaust gas as reduction gas on both side of the furnace arch, one side to the top of the main combustion chamber and the other to the flue gas, in order to reduce the NOx generated in the oxygen-enriched combustion.

The computational fluid dynamics technique is applied to simulate and analyze the performance of this new combustion system. A tow dimensional mathematical model has been developed to simulate the furnace structure design. Commercial software FLUENT has been used for the over bed coal volatiles combustion. Results show that the numerical simulation is successful and the pyrolysis and re-burning combinative technique generates a considerable improvement in efficiency for inferior coal combustion in modified grate coal boiler, and has good potential for future application in industry and for civil use.

AUTHOR INDEX

AUTHOR INDEX

AUTHOR INDEX

AIChE
120 Wall Street, FL 23
New York, NY 10005-4020

ISBN 978-1-5108-1250-5

CURRAN ASSOCIATES INC.
proceedings.com